Industry Analysis
The Telecommunications Industry

November 10–11, 1993
New York, New York

Randall S. Billingsley, CFA, *Moderator*
Richard S. Bodman
Stephanie Georges Comfort
Thomas W. Coyle, CFA
Lawrence J. Haverty, Jr., CFA
Barry A. Kaplan, CFA

Dennis H. Leibowitz
Maria F. Lewis, CFA
Mark Lowenstein
Robert B. Morris III, CFA
Charles W. Schelke, CFA
Raymond W. Smith

Edited by Randall S. Billingsley, CFA

To obtain an AIMR Publications Catalog or to order additional copies of this publication, turn to page 111 or contact:

AIMR Publications Sales Department
P.O. Box 7947
Charlottesville, VA 22906
U.S.A.
Telephone: 804/980-3647
Fax: 804/977-0350

The Association for Investment Management and Research comprises the Institute of Chartered Financial Analysts and the Financial Analysts Federation.

© 1994, Association for Investment Management and Research

All rights reserved. No part of this publication may be reproduced, stored in a retrieval system, or transmitted, in any form or by any means, electronic, mechanical, photocopying, recording, or otherwise, without the prior written permission of the copyright holder.

This publication is designed to provide accurate and authoritative information in regard to the subject matter covered. It is sold with the understanding that the publisher is not engaged in rendering legal, accounting, or other professional service. If legal advice or other expert assistance is required, the services of a competent professional should be sought.

ISBN 1-879087-31-6

Printed in the United States of America

May 1994

Table of Contents

Foreword v
    Dorothy C. Kelly

Biographies of Speakers vi

The Telecommunications Industry: An Overview 1
    Randall S. Billingsley, CFA

Global Telecommunications 8
    Robert B. Morris III, CFA

Understanding the Basics of the Local Telephone and Long-Distance Services Industries 15
    Charles W. Schelke, CFA

Internal and External Factors Affecting the Local and Long-Distance Telecommunications Industries . . 25
    Stephanie Georges Comfort

Local Telephone Companies 35
    Thomas W. Coyle, CFA

Competition in the Alternative Access Market 46
    Mark Lowenstein

The Cable Communications Industry 57
    Barry A. Kaplan, CFA

Strategic Issues in Multimedia: AT&T's Perspective 68
    Richard S. Bodman

Welcoming the Future: Bell Atlantic's Approach 73
    Raymond W. Smith

The Interactive Multimedia Market 79
    Lawrence J. Haverty, Jr., CFA

(continued on next page)

The U.S. Cellular Communications Industry Dennis H. Leibowitz	**85**
The Telecommunications Equipment Industry Maria F. Lewis, CFA	**97**
Glossary	**108**
Order Form	**111**
Selected AIMR Publications	**112**

Foreword

The most obvious characteristic of modern telecommunications is that the lines of distinction between cable, telephone, and computer services have blurred. For some time, the previously separate industries have been converging into a single telecommunications industry that is paving the information highway with a vast array of overlapping, fiercely competitive services and products. Opportunities change by the day as technological possibilities come to fruition and threaten to make today's products obsolete. Moreover, technology is opening communications among previously isolated areas of the globe. Innovations are, in general, transforming the speed, amount, and kind of communications we can attain.

Around the world, the demand for capital is growing as companies invest heavily in telecommunications infrastructure. In the United States, the competition arising from deregulation is spurring rapid growth in telecommunications services and a proliferation of public offerings. Furthermore, stocks that once traded like utilities are exhibiting the characteristics of growth stocks. Abroad, growth continues in the monopoly telecommunications providers, and privatizations offer investors surprising opportunities—but also the need to make careful judgments. For the investor in telecommunications stocks, the investment decision is thus increasingly complex. *The Telecommunications Industry* strives to clarify the investment choices and place the ongoing industry changes in perspective for securities analysts and others in the investment management profession.

The Telecommunications Industry is the sixth in AIMR's Industry Analysis series of seminars and proceedings. The series was conceived by Charles D. Ellis, CFA, to provide educational material on the nuances of individual industries that are relevant to securities analysis. This approach rests on the belief that the specific technical information that must be the backbone of any sound industry analysis is available only through personal experience with a particular industry. This series of seminars makes the fruits of that experience available to all.

The speakers at the seminar, whose presentations this volume reproduces, are among the leading specialists in the analysis of the telecommunications industry. AIMR wishes to thank them for sharing their research and practical experience and for assisting in the preparation of this proceedings. Special thanks are extended to Randall S. Billingsley, CFA, who moderated the seminar and contributed the overview that expertly sets the stage for the presentations that follow. AIMR also thanks John S. Bain, CFA, of Raymond James and Associates and John J. Keller of *The Wall Street Journal* for their help with the glossary, which begins on page 108.

The speakers contributing to the seminar were: Richard S. Bodman, AT&T; Stephanie Georges Comfort, Morgan Stanley & Company; Thomas W. Coyle, CFA, Coyle Research; Lawrence J. Haverty, Jr., CFA, State Street Research & Management Company; Barry A. Kaplan, CFA, Goldman, Sachs & Company; Dennis H. Leibowitz, Donaldson, Lufkin & Jenrette; Maria F. Lewis, CFA, Cowen & Company; Mark Lowenstein, the Yankee Group; Robert B. Morris III, CFA, Goldman, Sachs & Company; Charles W. Schelke, CFA, Smith Barney, Harris Upham & Company; and Raymond W. Smith, Bell Atlantic Corporation.

Dorothy C. Kelly
Assistant Vice President
Publications and Research
AIMR

Biographies of Speakers

Randall S. Billingsley, CFA, is associate professor of finance at Virginia Polytechnic Institute and State University. He has published articles in leading academic and practitioner journals. Dr. Billingsley has acted as a consultant to several large telecommunications firms and has served as an expert witness in public utility cost-of-equity determinations. He received a B.A. in economics from Texas Tech University and an M.S. in economics and a Ph.D. in finance from Texas A&M University.

Richard S. Bodman serves as senior vice president for corporate strategy, development, and intellectual property of American Telephone and Telegraph Corporation. He is a member of the Management Executive Committee and is chairman of AT&T Development Corporation. Mr. Bodman served previously as president of Washington National Investment Corporation and as president and CEO of Comsat General Corporation and president of Satellite Television Corporation, two major subsidiaries of Communications Satellite Corporation. He also served the U.S. government as assistant secretary of the Department of the Interior and as assistant director of the Office of Management and Budget in the Executive Office of the President. Mr. Bodman holds a B.S. in engineering from Princeton University and an M.S. in industrial management from the Massachusetts Institute of Technology.

Stephanie Georges Comfort is a principal of Morgan Stanley Company covering the telecommunications industry. She previously worked at Salomon Brothers, Gordan Capital, and Wertheim Schroder. Ms. Comfort graduated from Wellesley College with a B.A. in economics and has an executive M.B.A. from the Wharton School at the University of Pennsylvania.

Thomas W. Coyle, CFA, is president of Coyle Research, which publishes financial and market data on telecommunications companies. He was previously group vice president of Duff & Phelps Rating Company. He also worked for Ameritech Corporation in investor relations and in cost-of-capital analysis. Mr. Coyle received a B.A. in business administration from the University of Wisconsin and an M.B.A. from the University of Chicago.

Lawrence J. Haverty, Jr., CFA, senior vice president at State Street Research & Management Company, is also an equity analyst and member of State Street's Equity Investment Committee. Prior to joining State Street, he was senior vice president, director of research, and a security analyst at Putnam Management Company. Mr. Haverty received a B.A. and an M.A. from the University of Pennsylvania.

Barry A. Kaplan, CFA, is vice president and senior media analyst at Goldman, Sachs & Company, where he is responsible for coverage of the broadcasting, cable television, and cellular-telephone industries. Prior to joining Goldman, Sachs, he was an analyst at Bear Stearns & Company and at A.G. Becker. Mr. Kaplan graduated from Brandeis University and received his M.B.A. from the Wharton School at the University of Pennsylvania.

Dennis H. Leibowitz is a senior vice president/securities analyst at Donaldson, Lufkin & Jenrette covering the broadcasting, cable television, and cellular-telephone industries. Mr. Leibowitz is past president of the Media & Entertainment Analysts Society of New York, past president of the Cable Television Analysts' Group, and a past member of the board of directors of the International Radio & Television Society. He has a B.S. in economics from the Wharton School at the University of Pennsylvania.

Maria F. Lewis, CFA, is managing director at Cowen & Company, where she heads the telecommunications group and covers the telecommunications equipment and data-networking industries. Prior to joining Cowen, she worked for Lehman Brothers and Smith Barney Shearson, also covering the telecommunications and data communications fields. Ms. Lewis received a B.A. degree in economics from Rutgers College and an M.B.A. in finance from the Wharton School at the University of Pennsylvania.

Mark Lowenstein is associate director of the Yankee Group's wireless and mobile communications practice. His expertise is in cellular equipment, carrier services, competitive access providers, videoconferencing, voice-processing equipment and services, and international telecommunications. Prior to joining the Yankee Group, Mr. Lowenstein was a consultant with International Data Corporation. He holds a B.A. and an M.A. from the Fletcher School of Law and Diplomacy at Tufts University.

Robert B. Morris III, CFA, principal at Goldman, Sachs International, Limited, leads the global telecommunications services research effort. Prior to joining Goldman, Sachs, Mr. Morris was a securities analyst at Prudential-Bache Securities and a general partner at Montgomery Securities. He received a bachelor's degree and an M.B.A. from the University of New Mexico.

Charles W. Schelke, CFA, is managing director of the Research Division of Smith Barney Shearson. Previously, he worked as an investment analyst with Dean Witter Reynolds. Mr. Schelke is a past president of the Wall Street Utilities Group. He holds B.A. and M.B.A. degrees from the University of Washington.

Raymond W. Smith is chairman and CEO of Bell Atlantic Corporation. Previously, he held the titles of president and vice chairman. Mr. Smith currently serves on advisory boards of the U.S. House of Representatives subcommittee for renewing U.S. science policy, the Business Roundtable, and the Library of Congress. Mr. Smith is a trustee of the University of Pittsburgh and a member of the board of overseers of the School of Engineering and Applied Sciences at the University of Pennsylvania. He was the first recipient of the Mickey Leland Award for Diversity in Telecommunications from the National Association of Black Telecommunications Professionals. Mr. Smith is a graduate of Carnegie Mellon University and holds an M.B.A. from the University of Pittsburgh.

The Telecommunications Industry: An Overview

Randall S. Billingsley, CFA
Associate Professor of Finance
Virginia Polytechnic Institute and State University

The telecommunications industry has been among the best performing industries in the world in recent years. Yet, it can no longer be considered the safe domain of staid utilities operating in completely regulated, comfortable environments. Competition is here to stay—and to intensify—in all the various segments of the industry.

The changes in the telecommunications industry are driven by vigorous customer demand and fundamental changes in technology. The convergence of cable, television, telephone, and personal-computer use has created a dramatic tension between market and regulatory forces and a blurring of traditionally segmented lines of business. Interactive multimedia products are reported to be in demand even before the market seems to understand what they are, and strategic alliances to produce the products abound. Indeed, everyone seems in a hurry to get on the "Information Superhighway," although no one can say clearly where it is going.

What is clear is that technology and imperative market forces have forever changed the telecommunications industry and that investors must sort out the implications of the dramatic changes for both potential rewards and risks. The presentations in this proceedings help form a coherent framework for evaluating the often confusing, challenging, and potentially profitable companies in the telecommunications industry.

The industry is defined in this context to include companies providing local-exchange and long-distance telephone services, cable television (CATV), cellular (and wireless) communications, competitive access providers, interactive multimedia producers, and telecommunications equipment manufacturers. The discussions of each industry segment reveal the themes of increased diversification and integration among segments, the convergence of technologies, increasing competition, the tensions introduced by regulatory constraints and unequally yoked competitors, and the importance of distribution channels and the control of content.

A Global Perspective on Telecommunications

Robert Morris provides a global perspective on the performance of the telecommunications industry in recent years. He attributes the above-average performance of its equities to four key factors. First, telecommunications stocks have proved to be more growth stocks than utility stocks. In the majority of economies worldwide, telecommunications firms' unit growth rates have been between 1.5 and 3 times their country's underlying GDP rate.

Second, costs in the industry have been declining dramatically as capital is increasingly substituted for labor and the productivity of capital improves. In the telecommunications industry, once the formidable up-front capital has been spent, variable costs are negligible. Thus, the industry is impressively profitable because its asset-turnover rates continue to increase.

Third, unlike many other industries, telecommunications has not relied on pricing for its growth. Although declining prices in the industry are obviously not preferred by its companies, they do bring benefits. Declining prices have stimulated demand, especially in lesser developed countries (LDCs), and reduced competitive pressures.

Fourth, the rigid regulation of the telecommunications industry is softening throughout the world. Thus, new opportunities are being created for both existing companies and new market entrants.

Morris observes that telecommunications firms are unified by several common characteristics. Interestingly, the technology is fairly homogeneous worldwide. Furthermore, the distribution systems of various technologies are merging. For example, distribution systems of the CATV segment are merging with those of telephony, and the wireless distribution system is starting to resemble that of more conventional telecommunications. Digital technology is the primary facilitating factor. Service offerings also serve as a unifying characteristic of the industry; the majority of the vertical services offered throughout the world, such as 800 service, are similar.

The telecommunications industry is also characterized by a fairly consistent life cycle. The industry is in different phases of the cycle at different points of the globe. The early phase is associated with a large pent-up demand for services, lack of an established network, and the huge capital spending required by the need to lay cable and wires, install switching in the architecture, and build a network. Latin America is currently in this early phase. Companies start to reap the rewards of their investments in the adolescent phase. During the third, mature stage, competition reigns as firms cultivate the opportunities made available by their freedom from the sizable initial investments. North America is currently in a very late phase of the life cycle.

Regulation is also a unifying characteristic because it is fairly consistent throughout the telecommunications industry. Whether the regulation takes the form of traditional rate-of-return, incentive, or price caps, it almost always carries some social obligation. The obligation varies by country; it may be met by sharing profits with the public through pricing, for example, or through royalty payments made to the government, as in Hong Kong.

The unifying characteristics of the telecommunications industry provide the backdrop for valuation decisions. Morris emphasizes that valuation should be relative rather than absolute and that the focus should be on the relationship between total rates of return and relative P/Es. Supplementing this approach is the analysis of multiples of operating cash flow, as measured by earnings before interest, taxes, depreciation, and amortization (EBITDA). Morris offers provocative observations on valuing such telecommunications companies as AT&T, MCI Communications Corporation, and Telmex.

Local-Telephone and Long-Distance Services

Three speakers—Charles Schelke, Stephanie Comfort, and Thomas Coyle—provide insights into the basics, internal and external influences, and valuation issues associated with analyzing the local-telephone and long-distance services companies.

Schelke focuses on basics by describing the four historical phases in the life of the local-exchange and long-distance segments of the industry: development, stability and growth, transition, and competition. He views these segments as currently in a transition period from a highly structured, regulated, and monopolistic form to a completely deregulated, competitive structure driven by technological change. He predicts that telecommunications services will be in the full-competition phase by the end of this decade. Most regulatory restrictions, whether geographical or service restraints, will be eliminated. Some government regulation of pricing is expected to persist, however, especially for services to low-income, rural, or suburban users.

Furthermore, Schelke expects the long-distance, local-exchange, wireless, and cable industries to converge, with across-the-board competition among them. The seeds of this outcome are evidenced today by the energetic attempts to acquire, merge, or partner among telecommunications industry segments. These attempts include AT&T's desired acquisition of McCaw Cellular Communications, MCI's investment in Nextel Communications, and the various telephone–cable strategic alliances. Schelke predicts that every telecommunications service—broadband, cable, wireless, long distance, or local—will ultimately have multiple vendors. The technology will be fiber optic/coaxial cable and wireless.

Schelke argues that, as competition increases and opportunities as well as risks expand, companies seek to achieve three major objectives: (1) protection of as much of their own turf as possible and retention of market share; (2) entry into existing market segments not previously participated in because of regulatory or legal constraints (regional Bell operating companies [RBOCs] entering the cable or long-distance markets, for example, and cable moving into telecommunications); and (3) positioning to participate in new business opportunities, such as personal communication services (PCS) or multimedia services (for example, video-on-demand offerings, interactive games, or home shopping). New business opportunities are particularly attractive because they have the potential of significantly increasing revenues and cash flows over the companies' fixed-cost networks, which are already reasonably profitable in delivering traditional services. Wireless services in the United States are expected to grow 25–30 percent a year for the next ten years because current market penetration is only about 5 percent of the U.S. population.

In order to achieve these objectives, companies are adopting a variety of strategies. Some are moving up the value chain to actual or perceived proprietary products, such as obtaining brand names, increasing involvement in content, or developing proprietary operating systems. Partnerships with cable companies provide the opportunity to develop expertise on the cable/programming side and to acquire ownership or participation in content. Other current strategies are efforts to eliminate restrictive regulatory/legal barriers and traditional rate-of-return regulations; to enter foreign markets; to use partnerships, affiliations, and investments to penetrate nontraditional markets; to provide state-of-the-art services through digital switched broadband networks; and to become the lowest cost provider of services so as to discourage new entrants.

Schelke observes that the valuation of either local-exchange companies (LECs) or long-distance carriers in the new environment depends critically on four key operating issues. First, companies must determine what the proprietary aspects of telecommunications services will be in five or ten years. They will have to understand how to establish a proprietary position and maintain it. Second, given that the network companies generally have common technology, these companies will have to determine how they can differentiate themselves. Third, managers must analyze the current range of strategies to determine which are best and which they can adopt and implement at reasonable cost. Finally, the issue of whether technological trends will enhance or destroy the value of the existing network investments must be resolved.

Comfort discusses the core internal and external issues that are affecting the direction of local and long-distance telecommunications. She divides the external factors into five categories: regulation, competition, economic cycles, technological advances, and political stability. Regulation occurs on four levels: federal (through the Federal Communications Commission [FCC], state (local public utility or service commissions), the judiciary (exemplified by Judge Harold Greene, who handed down the initial consent decree responsible for the breakup of AT&T, administrative law judges, and district courts), and lawmakers (the U.S. Senate and House subcommittees).

As do other speakers, Comfort cites much evidence of the dramatic increase in competition in telecommunications. The landline network is being invaded by CATV companies and is threatened by the improving capabilities and increasing penetration of the wireless network. Although it is too soon to tell, the LECs could ultimately be allowed to enter the long-distance market as a quid pro quo for a competitive local loop. Furthermore, competition is coming from new entrants such as the competitive access providers (CAPs), which originally took advantage of opportunities to bypass the local-access business.

Comfort elaborates on the way economic cycles affect the telecommunications industry. She explains how inflation affects the cost of capital and how GDP growth is positively related to the growth in access lines for LECs and to the growth rate in minutes of use (volume) for the long-distance carriers. Figures for residential housing starts are often thought to be better predictors of industry growth, however, than GDP figures. As an external factor, advances in supplier and adjunct technology force changes in the telecommunications services companies. Comfort notes in closing her discussion of external factors that the globalization of the telecommunications industry increases the importance of issues related to the stability of foreign countries. Company managers must, at a minimum, assess the roles governments play in different countries, the currency issues, and companies' abilities to repatriate capital.

Companies providing local and long-distance phone services are affected by six basic internal industry factors: management, labor relations and costs, the state of the network, marketing strategies, diversification efforts, and financial capacity. Costs per access line and per employee have declined for all of the RBOCs during the past five years, and successful cost containment is expected to affect their future competitiveness markedly. Comfort observes that whether a company is bureaucratic (possibly because of the old "Ma Bell" orientation) or entrepreneurial can have implications for a company's flexibility in dealing with industry change.

Telecommunications managers characterize their diversification strategies in three ways: in-region versus out-of-region, international versus domestic, and vertical versus horizontal diversification. Comfort categorizes the approach taken by each of the RBOCs and by AT&T. In terms of average leverage and debt ratings, the LECs have the greatest degree of financial latitude; the cable and cellular companies are heavily leveraged.

Coyle concentrates on evaluating the fundamental characteristic of the LECs. He motivates his analysis by noting that the LECs provide the bulk of the RBOCs' revenues, operating profits, identifiable assets, and EBITDA. The major determinants of the parent companies' profitability are regulators, competition, diversification into new businesses, and growth opportunities in new services. In the new environment, a particularly important aspect for analysts is to evaluate the partnerships that will be formed and, in light of the amount of capital they continue to invest in their networks, the investment opportunities the RBOCs will have.

Coyle identifies key issues for investors and analysts seeking to value the telephone companies in an increasingly competitive environment. One issue is that the RBOCs have considerable investment in plant and equipment that are being depreciated with useful lives established by regulators. The important question for investors is how much of that investment the companies will completely recover before full-fledged competition emerges. Coyle also notes that responsibility for universal service is an issue of concern in the more competitive future telecommunications environment. New competitors—the CAPs, wireless carriers, and CATV companies—do not want the low-income Lifeline Residential Services customers. At some point, a battle over the allocation of costs for universal services is thus inevitable. Either the regulators will be under pressure to disallow some of the costs, or an LEC's competitors will find a way to get around them. Finally, Coyle notes that, as LECs move to cost-based pricing, they will shift cost burdens onto residential customers. The new competitors will eventually go after that residential customer base.

A major risk is that regulators will not give the LECs sufficient freedom to respond to competition. As competition increases, the LECs could face major asset write-downs as well as deteriorating profit margins and cash flows. The present companies' greatest risks reside with their diversification decisions.

Competition in the Alternative Access Market

Mark Lowenstein presents an overview of the important market for bypassing the traditional LECs to gain access to long-distance or interexchange carriers (IXCs). The CAPs usually have fiber-optic networks in major downtown or business/industrial parks. Their primary business is providing direct connections from the premises of large business customers to an IXC. When the CAPs initially entered the business, they offered 20–30 percent lower prices than the LECs for dedicated access lines. Many of the CAPs are small, privately held, or owned by cable companies. Although the CAP market is relatively small, it is expected to grow at a high rate in the near future because, as a result of the FCC ruling that the LECs must provide CAPs with switched access interconnections, the CAP segment can now compete in the switched-services market.

Among the factors that bring success to alternative access providers is being first to the market. CAPs have been most successful in markets where they are so far the only ones in the market, as in some secondary cities. The CAPs have not yet been very profitable, however, partly because of the RBOCs' pricing strategies in competitive zones and the relatively high interconnection rates charged to CAPs entering that segment.

Lowenstein provides numerous insights into the nature of the CAPs-versus-RBOCs competition and the relationship between CAPs and IXCs, cable, wireless, and PCS providers. For example, a potential problem for CAPs is that the IXCs are eager to build their own CAP networks.

Cable Communications

Barry Kaplan provides an overview of the fundamental issues affecting the cable industry in light of the convergence of the cable and telephone industries. He explains why cable companies rarely have earnings and why he believes that phenomenon is good. He also discusses how cable companies are valued, why telephone companies are trying to acquire or partner with cable companies, and what the successful cable company will look like in the future.

Kaplan argues that cable companies rarely have earnings because their historically entrepreneurial orientation sought to maximize tax-sheltered cash flow, net asset value, and overall cash return on equity rather than reported earnings. He views the lack of earnings as an advantage because the industry has never had to consider the impact on earnings of an investment decision, nor have the companies been concerned about dilution in an acquisition. The emphasis on cash flow has carried over to the cable analysts' valuation approaches. Most analysts and investors value cable companies based on cash flow, as measured by price-to-cash-flow multiples.

Kaplan suggests that the telephone companies' interests in acquiring or partnering with key cable companies stems from the explosion of enhanced services, such as home shopping and video games, that the telephone companies would like to offer but cannot because of numerous regulatory and physical obstacles. Cable companies offer a significant competitive threat to telephone companies' basic business, which telephone companies would like to offset by competing outside their own service areas. Acquiring or partnering with cable companies allows the telephone companies to expand beyond their geographical service areas. Furthermore, to the extent that such combinations bring programming assets, they enable the telephone companies to enter the content business, which is generally thought to have higher margins than distribution alone.

Cable is a business in which size is increasingly becoming critical, primarily because large size allows a cable company to have a say in how the technology evolves. Thus, Kaplan predicts that the successful cable companies of the future will be the "big" companies. In the past, "critical mass" in this industry segment was considered to be 1 million subscribers, but now it is on the order of 2 million subscribers. Large cable operators, such as Tele-Communications, Inc. (TCI), will have an enormous say in the evolution of the industry's technology and will thus be able to guide development in directions that are advantageous to them. Size also provides a company with the wherewithal to realize the large economies of scale possible in purchases of equipment and programming. Finally, the large companies will have access to the capital needed for future expensive network upgrades.

Strategic Issues for RBOCs and Long-Distance Carriers

Interesting insights into the strategic issues facing

managers of primarily telephony firms are provided by Richard Bodman of AT&T and Raymond Smith of Bell Atlantic Corporation. Their comments reflect the perspectives, respectively, of a major long-distance provider and one of the "Baby Bells" or regional Bell holding companies created by the AT&T divestiture in 1984. Bodman, as the key strategist for AT&T, predicates his ideas on the assumption that no single company has the capital necessary for the extensive plant and equipment of tomorrow's global telecommunications grid. AT&T will not necessarily own the network wires or wireless equipment, but as it does now, it will operate and/or administer an extensive global network and it will supply systems and hardware for networks built by other parties. In short, AT&T's strategic mission is simply to attract customers by extending and enhancing the capabilities of networks around the world.

Bodman describes AT&T's vision of the evolving multimedia landscape as based on the relationship between content providers, distribution companies, and consumers. On one end of the chain, the content providers offer entertainment, educational and training programs, games, information systems, software distribution, and shopping services. At the other end, consumers can choose many different access appliances—from television to wireless telephones, private branch exchanges, and local area networks. Connecting the two ends are the distribution companies, those in broadband interactive telecommunications, which includes cable companies, direct broadcast satellite companies, CAPs, LECs, and various wireless companies. The challenge is to make the instruments the customers use, through which the distributors provide access, as easy to use as a telephone. AT&T's strategy is to stimulate a market in which each of the delivery conduits can mimic or reproduce the capabilities of the others. Thus, AT&T would like to see all telephone companies, cable operators, and satellite groups have broadband, two-way interactive capability. Continued competition among these players will bring access costs down, which will help long-distance companies like AT&T deliver their services to customers.

AT&T views one of its greatest skills as that of integrating telecommunications plant, equipment, and services. One way to capitalize on this strength is to act as a host for content providers. The content products may be distributed to many different home receptacles, so AT&T hopes to offer the providers options—that is, networks—beyond consumers' TV sets through which to distribute their products. Potential additional services include the offering of content based on consumer selections and use and the ability to store content in digital form and to bill for that content as it is distributed.

AT&T's overriding strategy is to position itself as the best partner with which to build the infrastructure needed to support the multimedia landscape. Many companies are spending a lot of money to buy content, but whether they will have enough money left over to build a distribution system to deliver the acquired content to consumers remains to be seen.

Smith, CEO and chairman of the board of Bell Atlantic, discusses the strategic implications of the interactive multimedia phenomenon from the perspective of a Baby Bell. He indicates how Bell Atlantic hoped to exploit the convergence of the telephone, television, and the personal computer in the context of the then-pending but ultimately discontinued effort to merge Bell Atlantic and TCI, one of the major cable companies. Smith's worldview is summed up by three propositions: First, customer requirements have outgrown the boundaries of today's communications and entertainment businesses; second, the only effective response to the new customer requirements is rapid, industry-bending change; third, Bell Atlantic's transformation from a legal to a market franchise challenges investors to find new ways to evaluate and exploit the potential of what is, effectively, a new business.

Smith compares the video entertainment (cable) and communication (telephone) businesses in a way that explains their increasing convergence. Most of the cable industry consists of small, entrepreneurial companies that operate with minimal bureaucracies and modest capital. Their approach succeeds when growth can be achieved by adding subscribers or by simply buying more franchises. The approach falters, however, when competition presses a company to provide better customer service, to upgrade networks, and to decipher a more complex set of customer requirements than has ever been seen. The telephone companies, in contrast, have historically operated in an environment of rate-of-return regulation rather than competition. They are, therefore, large, bureaucratic organizations. They have generous capital bases that they tend to spend heavily, sometimes indiscriminately, on the basis of the perverse incentives created by traditional rate-of-return regulation. Their approach succeeds when the company is sheltered from competition but fails when competitors erode the company's best markets and take its best customers away.

Smith observes that both the cable and the telephone businesses have reached the limits of their original (legal) franchises and must look for new sources of growth. Both industries have mature technologies and customer bases, face competition from new entrants, and operate distribution systems that are limited in their ability to meet changing customer demands. Smith maintains that the two businesses can thrive only by reinventing themselves to meet the enormous market demand for interactive, on-line, personalized, easy-to-use, easy-to-find information, communications, and entertainment.

Bell Atlantic's strategic vision for its proposed merger with TCI was to extend and strengthen existing distribution franchises. Specifically, Bell Atlantic would invest in full-service networks that enable telephone platforms to generate cable and video revenues and enable cable platforms to generate telephone revenues. In addition, the planned merger was to expand Bell Atlantic's opportunities to participate in markets expected to be not only high growth but also to transcend geographical boundaries. In short, the primary goal of the planned merger was to establish a truly global franchise and brand identity in the markets for video programming, intelligent network capabilities, proprietary intellectual property, digital file servers, operating systems, and similar products and services. Consistent with the views of other speakers at the conference, Smith argues that the proper criterion for valuing such a new franchise is its effect on shareholders, as measured by cash flow growth and valued on the basis of a cash flow multiple.

The Interactive Multimedia Market

Lawrence Haverty, a buy-side analyst, takes a hard look at the optimistic projections currently being made for the size and profitability of the interactive multimedia market. He concludes that these projections reflect a great deal of fear, greed, and avarice. From an analyst's viewpoint, and in contrast to the viewpoint of Bell Atlantic's Smith, Haverty sees little that is new from this market for the investment business.

The multimedia segment consists essentially of two competing service-transportation mechanisms: cable and telephone. The manufacturers in multimedia create the content; they are such firms as Warner Brothers (films), Electronic Arts (video games), and PolyGram N.V. (records). The wholesalers package products manufactured by others; they are firms such as MTV (music) and Showtime and HBO & Company (filmed entertainment). The retailers include Blockbuster Entertainment Corporation, Warner Brothers Studio Stores, and the Disney Stores. The transportation networks link these elements.

Haverty stresses the necessity of a sound fundamental approach to valuing the equities of companies in the multimedia market. Following Warren Buffett's approach, he states that a good equity should possess four basic qualities. First, the business should be growing. Second, the business should be expected to generate cash some time during the investor's lifetime. Third, the business should have barriers to entry. Fourth, the business should have little or no government regulation. In applying that framework to multimedia's future, Haverty argues that the public is currently largely ambivalent about interactive multimedia and, therefore, the strong projected growth will take longer to materialize than is generally thought. Haverty believes, indeed, that certain signs indicate the multimedia market is a highly speculative environment. He notes a proliferation of seminars drawing big crowds, parties thrown by major players, evidence of a buildup in the inventories of those firms that have actually produced products to date, a growing number of companies that may well supply more product than the market will demand, and too many companies at which "chief scientist" is an executive position. Even with effective barriers to entry into multimedia, Haverty expects competition to be fierce as supply increases. Although his comments are partly tongue in cheek, Haverty makes the important point that the multimedia market may well be poised for some speculative excesses. Nevertheless, by exercising disciplined, fundamental analysis, investors can protect themselves from the crowd of speculators.

Cellular Communications in the United States

Dennis Leibowitz documents the phenomenal growth in the use of cellular communications, discusses the operating characteristics of the industry, outlines critical valuation factors, and predicts the likely future of the industry. He observes that market penetration has been remarkable but also notes that the performance of the stocks in the industry has been mixed.

The cellular business has healthy margins and is attractive economically. Once the hardware and the system have been established, incremental operating expenses are minimal; the flow-through of additional revenues to cash flow before marketing expenses can be as high as 80 percent. Controlling marketing costs is the key to meeting the competition's prices, and the cellular companies have successfully reduced marketing costs. Cellular service plans have increased margins and profitability by tailoring sales commissions and marketing costs to usage patterns. Capital investment per subscriber is also declining, because for most companies, the basic infrastructure is in place.

Leibowitz notes that two new kinds of cellular service are emerging to compete with the established wireline and nonwireline operators: enhanced specialized mobile radio and PCS, a miniaturized cellular system. In addition, cellular companies are expanding into paging and data transmission.

The valuation techniques most often applied to cellular companies rely on discounted cash flows of likely long-term performance. Initially, valuations of cellular companies were couched in terms of population. In order to find a per-pop value, analysts divide the present value of the interim cash flows and terminal value by the population of the franchise area. Leibowitz points out the pitfalls, however, in assuming that all pops are equal irrespective of particular markets. Thus, he notes, approaches to refine pops in relation to cash flows have been developed.

Telecommunications Equipment

Maria Lewis surveys the telecommunications equipment segment, which is growing rapidly in response to the cable and telephone segments' efforts to bring interactivity to consumers. She focuses on the telecommunications equipment manufacturers and vendors that supply infrastructure equipment to LECs and IXCs (a $70 to $80 billion market), cellular and PCS carriers (a $8 billion market), and CATV multi-system operators and other entertainment providers (a $2–$3 billion market).

Lewis divides the telecommunications equipment industry into a number of product–market segments. The major categories of equipment sold to carriers or service operators are central office switches, transmission and intelligent network equipment, wireless infrastructure, and CATV distribution plant and head-end gear. Excluding satellites and over-the-air broadcast systems, other product segments include equipment for business communications and customers' premises, data-networking equipment, and terminals or handsets.

The equipment industry's growth is fueled by increased customer demand, regulatory initiatives and outcomes; new carriers, services, and technologies; and the market opportunities presented by LDCs that are building infrastructure. Customer demand for new services is driven by a desire for business competitiveness, greater communications mobility, and increased personal productivity through modern communications and data-processing tools, as well as greater access to and diversity in entertainment and information sources.

As in other industry segments, deregulation has blurred equipment industry distinctions and is creating new operators. Internationally, privatization of many telecommunications entities is creating new capital and construction mandates. Spending on telecommunications equipment has also been boosted by the CAPs and data-only carriers that have risen to challenge cellular carriers. New technologies that have played a major role in the industry's accelerated development include digital switching, transport, processing, control, and compression; fiber optics and broadband wireless capability; and the rapid diffusion of intelligent network capabilities.

Competition among equipment suppliers is intense; barriers to entry—that is, substantial R&D and capital requirements and significant volume-based manufacturing efficiencies—are high and rising. Price considerations are often secondary to product differentiation; service; support; and in developing countries, financing, manufacturing, and job creation.

Lewis contends that the financial indicators to watch in the telecommunications equipment segment are much the same as those used by analysts for other manufacturing sectors, except that the need for R&D spending is greater than in many other industries. Of particular interest in equipment manufacturing are revenue growth, gross and operating profitability, leverage ratios, days receivables, and inventory turnover. Based on recent market results, telecommunications equipment stocks have high average multiples relative to data-networking shares even though, Lewis projects, near-term growth rates are likely to be higher for the latter.

Global Telecommunications

Robert B. Morris III, CFA
Partner
Goldman, Sachs & Company

> Although telecom companies around the world are in various stages of development and operate in different economic environments, they also have a number of common characteristics. In general, their stocks perform well in comparison with those of other industries.

The global telecommunications industry is one of the best performing industries in the world. It has consistently beaten the broad market averages. Telecommunications stocks around the world have performed well on a capital-appreciation basis during the past five years. In addition, most of the stocks in this universe pay healthy dividends and have yields that are superior to the overall markets in which they participate.

Managers who invest money globally often compare their global performance against the FT-Actuaries World Index. In both local currency and U.S. dollar terms, investors will find the telecommunications industry in general, and specifically the worldwide telecommunications industry, worth considering.

The Strengths of Telecommunications

Four key factors explain why telecommunications is worthy of an investor's time and consideration.

■ *Growth.* Historically, many investors viewed telecommunication companies as utilities rather than growth companies. Today, however, telecom stocks are growth stocks, and they will be growth stocks for the next decade. The telecom industry is an industry with underlying unit growth. It is not like some industries that were considered growth industries in the 1980s but did not have true underlying unit growth characteristics on a global basis. In almost all worldwide economies, the telecom companies have sustained their unit growth rates for five- to ten-year periods, and these growth rates have been 1.5 to 3 times the underlying GDP growth rates of their home economies.

■ *Declining costs.* Costs in the telecom industry are declining dramatically in two respects. One is the widespread substitution of capital for labor. The head counts of telecom companies have been dwindling dramatically for the past seven years, and this decline shows no signs of abatement. Telecom Corporation of New Zealand, for example, by substituting different types of capital for labor, now serves more than 180 telephone access lines per employee; it represents the high end of the range. About eight months ago, it announced a further restructuring program. By substituting capital for labor during the next three years, the company hopes to reach its objective of more than 300 access lines per employee.

The second aspect of declining costs is productivity of capital. Telecom differs from the airline, power, and other industries that have huge up-front capital costs and high variable costs. The power industry, for example, requires tremendous up-front investment to put the distribution network in place. In addition, it has high variable costs related to the source of power.

In the telecommunications industry, once the capital is in place, the variable costs are almost zero. In an unregulated environment, telecom is a profitable industry with tremendous capabilities for generating cash flow. For example, Hong Kong Telecommunications Ltd. and Singapore Telecom are not regulated in the traditional sense. They have no debt in their capital structures. They have payout ratios as high as 70 percent. They generate excess cash, and their return on equity is between 40 and 50 percent in an inflationary environment of 2–3 percent. Telecom is an immensely profitable industry because asset turnover rates are continually escalating.

■ *Price elasticity.* Unlike other industries, telecom has never relied heavily on pricing for growth. About 18 months ago, investors characterized the tobacco industry as a growth sector. To get that growth, Philip Morris Companies and other companies relied primarily on price. Most telecom companies have been subject to continually declining prices. Although they would prefer not to have declining prices, declining prices hold two benefits for telecom. First, they stimulate demand, particularly in lesser developed countries. British Telecommunications PLC (British Telecom), for example, occasionally runs specials on Sundays. It will drop prices 20 percent for a day and advertise the price break heavily. Volume grows at least 20 percent on that day. The elasticity factor during that period is 1, and no substitution effect occurs—that is, customers do not advance a call they would have made later in the week. The market is price elastic, particularly in certain segments such as international long distance and in lesser developed countries with pent-up demand. Although price stimulates demand and true unit growth, telecom has not relied on price and will not in the future. Second, declining costs reduce competitive pressures.

Deregulation. Severe regulation of the telecom industry is easing around the world, and deregulation is creating new opportunities both for entrenched operators and for new smaller entrants. For investors, those opportunities translate into the opportunity to make money in stocks.

Analytical Framework

Four major characteristics unite the telecom industry across borders: homogeneous technology, similar service offerings, a predictable life cycle, and consistent regulation.

Homogeneous Technology

The technology of telecommunications is fairly homogeneous worldwide. Telecom is unique because the operator does not have to develop the technology. It does not necessarily develop the product. Sometimes the companies developing the technology develop new applications for telecom companies' networks. For example, telecom companies such as AT&T, BellSouth Corporation, or British Telecom may not have developed the fax machine, but their networks benefitted from its development far more than any of the fax machine manufacturers. For the telecom networks, fax business is a continuing annuity.

A standard technology creates global ubiquity, which is a great benefit to the telecom sector. This standardization allows for the development of hardware that can be used throughout the world, including modems and communications for personal computers worldwide. My laptop computer, for example, has two plugs: one for use in North America and one for use in Europe. I simply plug it in and communicate directly with other computer operators.

A collapse is developing on the technological front of the telecom industry. It is most pronounced in the United States but is beginning to accelerate throughout the world. The distribution systems of various technologies are beginning to merge. The distribution system of the cable TV system is merging with and looks a lot like the distribution system and the technology of the telephony distribution systems. The wireless distribution system is beginning to look like the distribution systems of telecommunication systems. Digital technology allows this to happen. When machines start communicating with each other digitally, the process is easier.

The hierarchies of the switching networks are flattening, just as the hierarchies of the old computing networks are flattening. The old computing networks had centralized information distribution and control of information in the mainframe computing environment. The distributed-processing client server has changed the environment. Some upstart companies and some older companies such as AT&T are creating overlay networks on top of old networks, which followed the distributed-processing client server environment. The changing hierarchies allow for rapid product introduction and more standardized product introduction across the network at a given time.

Service Offerings

Service offerings are a second unifying characteristic of the industry. All of the vertical services, such as 800 service, that large operators offer are similar. The service offerings and economic characteristics are the same all over the world. If investors understand the markets in the United States, they can look at global telecommunications companies and glean some insights on how those markets and earnings might behave and the leverage the introduction of a new service offering provides.

Life Cycle

The telecom industry has a fairly consistent life cycle, which serves as a third unifying characteristic. Investors can see different phases of the life cycle around the globe. Latin America is in a very early phase, and North America is in a very late phase. Typically, the early phase is characterized by huge pent-up demand for services, lack of an established network, and tremendous capital spending requirements based on the need to lay cable and wires, install switching in the architecture, and build a network.

During the adolescent phase that follows, companies realize some benefits from the investment. During the third and more mature phase, competition enters the market, and other opportunities exist because companies no longer need to make huge initial investments. They have an opportunity to develop operating characteristics, particularly the asset turnover characteristics.

Regulation

Regulation is also fairly consistent throughout the telecom industry. Regardless of the form regulation takes—price caps, rate of return, or incentive regulation—providing telecommunication services includes a social obligation. Social obligations can be met in many ways, but many societies believe that the riches of the industry must be shared with the public whether through pricing or, as in Hong Kong, through royalty payments paid to the government.

In many cases, the large, entrenched basic service operators share the riches of the industry with the broader public. Companies in Thailand have BTO (build, transfer, and operate) licenses. For example, Shinawatra, which has a license to build a million lines outside Bangkok, installs the lines, builds the system, transfers it back to the authorities, and receives a royalty from it.

Relative Valuation

Investors can analyze common themes in this industry to determine relative valuations and make valuation decisions. A broad, consistent, global approach examines several characteristics. Investors should be aware that no absolutes exist. Just because Telefonos de Mexico SA (Telmex) trades at 18 times earnings does not mean it is cheap, and just because Singapore Telecom trades at 50 times earnings does not mean it is expensive. The valuation depends on the operating environment.

Many factors affect absolute valuation levels. Mexico, for example, has an inflation rate of about 10 percent. Its sociopolitical environment differs from that of Singapore. The regulatory situation may not be as transparent in Mexico as it is in Singapore. Singapore has had inflation of about 2 percent, real GDP growth of 8–10 percent, and consistent currency appreciation of about 6 percent for seven years. All the investment factors are transparent. Investors know where the politicians stand, how the government works, and how the company works. They know if the company or government says something, that is what will happen. Low risk premiums are built into the overall valuations in the market resulting in higher absolute price to earnings ratios (P/Es) than, say, in Mexico.

Investors should also examine relative total rates of return. For example, Hong Kong Telecom enjoys growth of 13–15 percent a year. Dividends consistently grow at about 15 percent a year. It is operating in an inflationary environment of 10 percent, so its real earnings are growing at 5 percent. In general, earnings in the Hong Kong market grow about 16–17 percent a year. The yield on the market matches that of Hong Kong Telecom's stock. Thus, the total rate of return for Hong Kong Telecom is less than that of the overall Hong Kong market.

To find value, however, investors should analyze the relationship of total return to relative P/E. Hong Kong Telecom currently trades at about 23 times earnings, and the Hong Kong market trades at about 15 times earnings. The difference works out to almost a 50 percent premium to the overall market. Hong Kong Telecom's total return is about a 20 percent discount to the overall market. The stock is trading at 150 percent and returning 80 percent of the overall market. Based on a simple valuation test (relative total return divided by relative P/E), Hong Kong Telecom is overvalued. Investors are getting no value for their money and are paying dearly for growth and total returns that are far less than those of the overall market.

Value, stock outperformance, and relative valuation are related, although not exactly in a linear fashion. Companies such as Compania de Telefonos de Chile (CTC); Telmex; Sta Italiana per l'Esereizio delle Telecommunicazioni (SIP), the Italian telephone company; and Sta Finanziaria Telefonica Torino (STET), a holding company that owns the local Italian telephone company and the international long-distance operator, provide total rates of return in excess of their overall markets. STET, for example, trades at a valuation less than that of its overall market.

Investors are getting very good value for their money when they buy CTC, which has grown faster than the overall market and offers good total rates of return. After adjusting for inflation, its average growth rate is about 23 percent a year; the overall market is growing at about 15 percent. In nominal terms, CTC grows at about 36 percent and yields about 2 percent. Even though the stock offers premiums of about 30 percent to the market in growth and total returns, it trades at a market multiple.

This relative valuation analysis does not always work in the short term. For example, in June 1989, stocks in Hong Kong became cheap. We recommended purchase of Hong Kong Telecom in July and held the stock until June 1993, when the analysis indicated that it was overvalued. During September and October, however, a flood of money moved from North America into the Hong Kong market. Many

North American investors like to buy American Depositary Receipts (ADRs), and Hong Kong Telecom is one of the few ADRs available for this market. Chinese Investment Trust (CITIC) owns 15 percent of Hong Kong Telecom and is rumored to be a seller of the stock. CITIC is a quasi-governmental entity of the Chinese government and an insider and board member of Hong Kong Telecom. Hong Kong investors are selling the stock, but the weight of foreign money has pushed the stock up some 14 points since the end of June 1993.

In the short term, some developments may work against telecommunications investors, but in the long term, analyzing value, stock outperformance, and relative valuation will lead to superior performance. Investors can beat the markets by analyzing stocks this way.

Nevertheless, investing in global telecommunication stocks does entail an incremental level of risk because of the volatility of currencies. For example, British Telecom was up more than 25 percent in 1992, but it was flat in the United States because the pound plunged when the United Kingdom pulled out of the Exchange Rate Mechanism. The returns for U.S. dollar-based investors were essentially eliminated by the currency fluctuation. Investing offshore is subject to periods of currency and political gyrations. Managers who are measured against a U.S.-based index should tell clients up front that currency management is not their area of specialty and that they are making currency-neutral calls.

Alternative Valuation Methods

The investment characteristics of the telecom industry are changing, particularly in the United States, where the telecom industry is shifting its investment patterns. These changing investment characteristics are forcing analysts to examine other valuation approaches.

In 1992, Bell Atlantic Corporation spent about $300 million on copper wire to upgrade its network (also referred to as outside plant). In the spring of 1993, U S West announced it would rebuild its network. During the summer, it took a $5 billion pretax write-off of its outside plant. These companies were investing in an asset that has very little or no commercial value the minute it goes in the ground. Now, the industry is beginning to change how it makes its investments. For example, U S West, with an overbuild of its network, is trying to construct a network in Omaha that generates not only basic telephone services but also multimedia services.

The changing investment characteristics in this industry—spending in new areas and shortening depreciation lives to reflect the economic realities of competition—are putting pressures on the industry. This change forces analysts to turn to valuation methods such as multiples of operating cash flow (defined as earnings before interest, taxes, depreciation, and amortization—EBITDA) and stock values per access line. Investors will find significant differences among the leveraged cash flow multiples of telecom companies. This multiple represents the total value of the firm (debt plus equity) divided by operating cash flow. Hong Kong Telecom, for example, at $48 a share, trades at about 16 times cash flow and $7,000 per telephone access line. In November, the stock was trading at about $61, or close to $10,000 per access line and 20 times operating cash flow. Economically, revenues that justify, on a present value basis, a $10,000 value per access line are difficult to generate. Investors must consider what types of revenue streams and what types of profits must be generated on a discounted cash flow model to yield such high stock values. To justify current prices, earnings must grow 14 percent a year between now and 2010 and then 8.5 percent a year into perpetuity. China is a great opportunity, but increasing telephone revenues and earnings at 8.5 percent into perpetuity means not only that everyone in China has a telephone from Hong Kong Telecom but also that everyone in the world has more than one telephone from Hong Kong Telecom.

In contrast, some stocks trade at very low cash flow multiples. In the United States, for example, some companies are trading at a little more than 7 times cash flow. By expanding into commercial environments in which they have more control, these companies have the opportunity to expand their cash flow multiples. That is why, when Bell Atlantic announced the proposed merger with Tele-Communications (TCI) and the fact that this merger produced 35 percent earnings dilution, its stock price increased. Our firm recommended the stock that day following a major internal debate. A codirector of research questioned how the stock of a company with 35 percent earnings dilution could increase in value. Others in the firm were interested in the $5 billion in cash flow TCI would contribute, how that cash flow would be used, what changes might occur, and where the support for such valuations might lie. They concluded additional value might exist. I compared the new company to Telecom Corporation of New Zealand, which competes in the marketplace today, trades at about 8.5 times cash flow, is not restrained by many of the characteristics of the U.S. market, and can do anything it wants.

Analysts should temper how they look at valuations by other characteristics such as cash flow multiples, because the industry's investment characteristics and depreciation characteristics are changing. The equipment going into the ground today differs from what went into the ground five years ago, particularly in the more well-developed economies of the world.

Application of Analysis

As an example of how this broader approach is applied, I will examine some global telecom stocks and explain ways to look at them, opportunities in the stocks, and the thought process behind the conclusions.

British Telecom, unlike its counterparts in the United States, is fully integrated. It is the international long-distance carrier, domestic long-distance carrier, and local service provider in the United Kingdom. Relative value is a major issue for British Telecom. Its fundamental position is improving. The industry has significantly upgraded its technology so it is almost state of the art. The regulatory environment has been changing for the better, and developments on the competitive front have also worked out for the better. These conditions lead to opportunities for margin expansion, an outgrowth of the asset turnover rates and the asset turnover opportunities for this industry. Margin expansion, in turn, will lead to greater profitability and an opportunity for improved revenue and earnings growth.

In contrast, the Spanish telephone company, Telefonica, like many of its counterparts, is a play on the local stock market. Telecom stocks tend to be the primary vehicles for global investment. They tend to be the global investor's first foot into a market and the biggest, most liquid stock in a market. Because of the global unifying characteristics of this industry, investors can cross borders and understand what the companies are doing in various markets. The Spanish telephone company is much more heavily leveraged than most companies around the world.

The earnings outlook for AT&T is improving as a result of rising productivity and declining costs, which provide opportunities for margin expansion and improved profitability. Investors should also consider the issue of strategy. AT&T, like a combined U S West–Time Warner, is very well positioned for some of the coming changes in the U.S. market.

Investors will find opportunities in the big operators, but they will also find opportunities in the little operators that are picking off segments of the market. For instance, ALC Communications is a great company emerging from a recent restructuring. It aims at a very narrow niche within the long-distance sector, and it does it very well. The company addresses a very narrow segment—the commercial business segment—with a very motivated, young sales force.

Although AT&T and MCI Communications Corporation spend hundreds of millions of dollars trying to convince customers that they have the truest sound and that the network can be differentiated, when a customer makes a long-distance call, he or she would have difficulty figuring out who the carrier is. In the long-distance segment, technology is not a differentiating characteristic but a uniting characteristic; marketing is the differentiating characteristic.

Telmex has very attractive valuations on a relative valuation basis. The company is in the early stage of its life cycle. It has attractive underlying growth characteristics, pent-up demand, and good opportunity for improving productivity trends. Because of the passage of the North American Free Trade Agreement (NAFTA), I expect the stock will increase handsomely.

Chile's two telephone companies, Compania de Telefonos de Chile SA and Entel Chile are in the early phase of development, and they should mimic Telmex's early performance.

In New Zealand, investors will find an improving fundamental situation, good valuation, and dramatic cost reductions. Telecom Corporation of New Zealand can be a model for U.S. companies for valuation, opportunity, cash flow, and growth. I would encourage investors to look at it for comparison purposes.

Conclusion

In conclusion, I would like to repeat my initial point that the global telecommunications industry is one of the best performing industries in the world. With careful analysis of some of the factors I have outlined, investors should find good returns from many of the stocks in this sector.

Question and Answer Session

Robert B. Morris III, CFA

Question: Given that relative valuation versus local markets makes sense, how do you choose among markets? Can a firm be overvalued in its local market and still be attractive from another market's perspective?

Morris: Ultimately, I think not, but in a rising tide, all ships go up. Extreme bullishness about the Hong Kong market spurred the appreciation in Hong Kong Telecom. Telecom stocks tend to be the way global investors play a foreign market, but those plays usually come back to haunt them. When a market backs off or corrects, the most expensive stocks in the market will be hit hardest. Investors who use a particular stock as a proxy may suffer when the world decides to sell that market. For example, when the world decided to sell Mexico, the price of Telmex plummeted.

Telefonica de España is a telling example of how to look within a market. In the 1980s, Goldman Sachs did an underwriting for Telefonica. It looked like a cheap stock to Americans. Based on U.S. GAAP, it was very cheap—6 times earnings. U.S. GAAP, however, is not particularly relevant when investing in other markets. Under Spanish GAAP, Telefonica was trading at about 14 times earnings in a 12-P/E market. For five years, the stock did nothing because the Spaniards were convinced that, if U.S. investors were dumb enough to want the stock, they could have it all. The Spaniards kept shoveling it out the door until a fundamental change occurred in Spain, and they decided the stock was worth more. The stock then moved up.

Question: What is the future of Telmex? What impact will NAFTA have?

Morris: Telmex was weak in anticipation of NAFTA failing and was up slightly after passage. Telecom is a soft cyclical industry. It is sensitive to the economy, although it still has unit rates of growth. During the current recession, the overall domestic U.S. long-distance sector has shown unit growth rates of 5–6 percent in a contracting economy. Although that is still growth, it is a materially lower growth rate than in a growing economy.

Mexico has been going through a contraction, and the underlying rates of growth are slowing. The effect is more pronounced on a company with a large capital expenditure program that is bulking up and building out. Slowdowns in revenues and rates of growth in revenues associated with the economy are usually more severe in expanding companies. Simultaneously, unit growth is slowing, the base is growing, and the combination pinches the company's margins.

My expectation is that a little revenue growth will go a long way. For example, British Telecom reported its second-quarter earnings the other day. The economy is picking up in the United Kingdom. Revenues were maybe 1–2 percent higher than consensus forecasts. Pretax operating income before redundancy charges was 10 percent higher than consensus projections.

Question: What do you think of Telefonica de Argentina versus Telecom de Argentina?

Morris: Telefonica de Argentina is partly owned by Telefonica de España, France Telecom, and STET. If you know the parents, you know the children. The strategies are consistent with what they have done in their own markets and what they are doing elsewhere. France Telecom went through a tremendous system upgrade, has a very modern and efficient network, a real focus on quality, and a balance between building a network and maintaining quality. Its relative, Telefonica de España is doing the same but is showing better earnings growth because it is exploiting the short-term opportunity of building out an underdeveloped system.

Telecom de Argentina is growing, but it has accelerated its depreciation. It is writing off plant and equipment very rapidly. It is trying to modernize the existing network rapidly, so earnings growth rates are somewhat slower than Telefonica de Argentina's. A couple of years from now, that situation will reverse. Compared with Telefonica, Telecom de Argentina's investment in depreciation and modernization of the network will pay tremendous dividends.

The big risk for many of these companies is the social obligation. Telefonica ran into problems because of its service quality. It was building its network but not keeping its eye on its customer base. As a result, it got in trouble with the authorities. That can happen very easily in Argentina.

Question: What U.S.-based companies will gain most from your bullish outlook for global telecommunications?

Morris: The best positioned industry in the United States is the long-haul industry—the AT&Ts and MCIs. The merging of local distribution technologies will create competition for local distribution and drive prices down. Declining prices will stimulate demand, which will contribute to unit growth, particularly because companies are increasingly migrating to broadband networks.

I envision high demand for new value-added products and the increased growth in units aiding the long-haul industry. For example, Sega Games will not distribute its game software in every city and town in the country. The big expense for video distribution systems will be the servers and storage devices, which will be centralized devices. The video will travel over long-haul networks. The growth should accrue to the long-haul networks. The other self-evident reason is that increased competition in local distribution areas should bring prices down. AT&T is in the best position because of its labs, manufacturing ability, and scale. Part of the reason MCI sold a stake in itself to British Telecom was to gain scale. To play at that level takes tremendous resources. AT&T already has the scale.

Question: What is the potential for investment in India?

Morris: It is a country play. The world is moving in a different direction than the United States. Many Southeast Asian countries are embracing the capitalist society and flourishing of economic growth. India is on that train. One of the real beneficiaries of that is the telecommunications sector.

Understanding the Basics of the Local Telephone and Long-Distance Services Industries

Charles W. Schelke, CFA
Managing Director
Smith Barney Shearson

> The telecom industry is moving through a period of transition to a more competitive environment. Companies are using different strategies to meet the changing industry structure and dynamics.

The structure of the telecommunications services industry is changing dramatically, and companies are responding to the changes in a variety of ways. To understand the basics of the industry, investors and analysts should start with its history and characteristics.

Historical Phases

The history of the telephone (also called local-exchange companies, or LECs) and long-distance industries can be divided into four phases: development, stability and growth, transition, and competition. The first phase lasted from about 1890 to the 1920s. During this period, Alexander Graham Bell invented the telephone; AT&T was organized; and following the expiration of the Bell patents, independent telephone companies emerged, typically in small rural or suburban areas where AT&T did not operate. During this time, the independents and AT&T went through the same confrontational process as seen today with new entrants over the right to interconnect. Under the threat of an antitrust suit, AT&T agreed to work in cooperation with the companies that were operating outside of AT&T territories.

The second phase, which began in the 1920s, was a period characterized by stability and growth in the industry. It operated as a well-defined partnership. Each company served a defined franchise, or geographic territory, interconnecting through AT&T's long-distance network. AT&T set the standards as the industry went through the process of wiring the country.

The industry is now in a transition period, moving from a highly structured and regulated, monopolistic industry to a completely deregulated, competitive structure driven by technological change. This phase can be dated either from 1968, when MCI Communications Corporation was organized to provide long-distance competition, or from 1983, when the Bell system broke up. This phase should continue until about the year 2000.

Starting near the end of the decade, telecommunications services will be in the full-competition period. Most regulatory restrictions on the industry, whether geographic or service restraints, will be eliminated, although some light government regulation of pricing will persist, particularly for low-income, rural, or suburban telecommunication services users.

The industry has four major sectors—long distance, local exchange, wireless, and cable—which aggregate about $170 billion in revenues. Today, the sectors have some slight competitive overlap. For example, in some states, long-distance companies compete with local-exchange companies in local toll, and some modest competition exists between cable and telcos. Little competition exists between the wireless (cellular) companies and the LECs, however, and most wireless calls continue to go through the LECs.

By the year 2000, all of these sectors will converge, and the industry will have across-the-board competition. For example, the telephone companies will become broadband companies providing entertainment just as cable companies do. The regional Bell operating companies (RBOCs), which are now prohibited from providing long-distance service, will be involved in regional, national, and international long-distance service and compete with the existing long-distance companies. The long-distance companies will move into the local-exchange business through a combination of both wireless and wireline facilities, either owned directly or in part or resold. For example, as LEC networks are opened up, long-distance companies could install their own switches, lease transport capacity from the LECs and position themselves, from a customer standpoint, as local carriers. Cable companies will be upgrading their systems from a broadcast to a switched mode and providing two-way telecommunications services.

All of these elements are being put into place as evidenced by the acquisition/partnership activity that has been going on. This activity includes AT&T's proposed acquisition of McCaw Cellular Communications, MCI's investment in Nextel, and the various telephone–cable alliances. The rapidity with which this convergence process is happening is striking, and there is little doubt that, in a relatively short period of time, many users will have a choice among wireline broadband carriers, as they do now among long-distance companies, and to a lesser but increasing extent wireless carriers. Thus, by the end of the decade, competition in what has been a monopolistically served industry will be extensive.

Industry Structure

The primary forces driving this structural change have been technology and consumer desire for alternative vendors. This change has also been accelerated by the value-of-service/universal-service rate structure that has historically been used in telecommunications. This non-cost-based rate structure has enabled new entrants to expand relatively rapidly and profitably in market segments such as long-distance and business services that historically have been overpriced to subsidize residential service. Even without the rate structure advantages (to the new entrants), however, technology would have forced structural change in the industry.

Technology

Telecommunications technology was fairly stable during the period from the 1890s to the early 1970s. The technology was copper and electrical/mechanical based, and telephone engineers of the 1920s would not have been surprised or confused by the technology of the 1970s.

The technology began to change dramatically in the 1980s, moving from analog to digital, from copper to fiber, and from a combination of electrical/mechanical to optical/electronic and wireless technology—a whole new way of providing communication services. The new technology has had and will continue to have profound implications for the industry, because it dramatically lowers the cost of providing services and expands the array of services that can be offered. In essence, capital costs in telecom are following the same trends as in the computer industry with price–performance ratios doubling about every 12–18 months.

Electronic/optic technology also has considerably lower operating and maintenance costs and provides dramatically greater reliability than copper/mechanical technology. During the past ten years, most telephone companies have reduced their labor forces by about one-third. In terms of the operation of the network itself, however, labor force reductions have been close to two-thirds, resulting in a dramatic change in the industry's operating cost structure.

New technologies also enable companies to provide new software-driven services at reasonable prices. These services include custom billing, Caller ID, virtual private networks that permit a company to operate a private network using the public switched network system, and high-speed data transport. These advances are all moving toward interactive broadband switched networks that will enable consumers, at reasonable cost, to call up almost any entertainment or data service they want whenever they want it.

Regulation

Historically, the telecommunications industry has been regulated by federal and state authorities as utilities on a rate-of-return basis. That system provided strong industry incentives to maintain a relatively high rate base and a high cost structure. Also, since universal service was a stated goal of regulation, rate structures evolved that subsidized residential users through toll and business services. Thus, rates in many sectors have been unrelated to costs.

The most dramatic case of overpricing to support local service occurred in the long-distance industry, particularly after World War II, when costs dropped dramatically as a result of the use of microwave radio, which was commercialized during the war. The industry allocated costs to the long-distance sector to avoid raising prices for local service, the costs of which were trending up. Although telecommunication costs are inversely proportional to density and use (customers in large cities are cheaper to serve than those in small rural areas, and large customers are cheaper to serve than small ones), large customers paid as much as or more than small customers on a unit-volume basis and rural customers often paid less than urban customers. This skewed rate structure with its artificially high long-distance rates enticed entrepreneurs such as Bill McGowan (MCI) to enter the long-distance business. It was a major factor in the development of a competitive long-distance industry, at least in the early years.

As competition has developed in long distance, most of the margin differential between customer groups has been adjusted, with some major exceptions such as international long distance. Nevertheless, the rate structure in the telephone industry is still oriented toward value of service rather than cost of service. Thus, sectors such as access (the origination or distribution of long-distance traffic within telco service territories) are still priced far above costs. Pricing is a major issue that the telephone industry (defined to include regulators) must face in moving to a competitive environment. The transition will not be all that difficult, because the adjustments required typically amount to about $5–$10 a month per customer; if basic rates are raised $5–$10 a month, toll and access rates can be cut by 50 percent or more to bring them closer to the actual costs of providing services.

By the end of this decade, the industry is likely to have few if any regulatory constraints that limit the ability of individual carriers to provide services from a sectoral or geographic standpoint. Ironically, this prospect means that 15–20 years after the breakup of the Bell System eliminated one-stop shopping for telecommunications services, the change to a competitively structured industry will probably bring about a return to a single-vendor supplier of expanded services for many customers. In contrast to predivestiture days, however, when there was only one vendor—the local telephone company—tomorrow's environment will allow consumers to select among several vendors to provide some or all of their telecommunications/entertainment services.

Although the industry is moving away from rate-of-return regulations, some regulatory oversight will persist, along with certain subsidies to provide telephone service for high-cost areas or low-income consumers.

Summary

To summarize all the above, the telecommunications industry is likely to be almost totally deregulated by the end of the decade. Every telecommunications service—broadband, cable, wireless, long distance, or local—will have multiple vendors. The technology will be digital; transmission will be fiber optic/coax cable and wireless. The system will begin moving toward all-optical technology. The market will set prices, primarily based on the cost of providing services, although companies will try to provide actual or perceived proprietary products and services that they can price well above costs. Capital cost trends will continue to decline, and the capabilities of the system will expand rapidly.

These developments are fairly dramatic. The long-distance industry has already gone through the process, which took about 15 years. Because companies and regulators have already been through interconnection and other issues once, the next phase (local-exchange competition) should be resolved much faster.

Company Goals and Strategies

As competition increases and opportunities (and risks) expand, companies seek strategies to achieve three major objectives:

- Protect as much of their own turf as possible and retain market share.
- Enter existing market segments that they have not participated in because of regulatory, legal, or other reasons. Examples are the RBOCs entering the cable or long-distance markets and cable companies moving into telecommunications.
- Position themselves to participate in new business opportunities such as multimedia services (video-on-demand, interactive games, or home shopping, for example) or PCS.

New business opportunities are particularly enticing because they offer the potential of significantly increasing revenues and cash flow over a fixed-cost network that is already reasonably profitable on traditional services. The mathematics of this potential is as follows: The average family spends $60–$65 a month for telephone and cable service combined—$25–$30 for telephone and $30–$35 for cable. (Cash flow from these combined services is about $25–$30 a month.) If companies can expand that total by $10–$30 a month at small incremental cost, profitability and cash flow could increase by 25–50 percent.

Multimedia services are expected to be highly successful (although the time frame is uncertain) as communications companies move into existing markets served by other industries. Video-on-demand, for example, will compete with the $15 billion a year market for video rentals. Home shopping could displace some or most of the $50 billion catalog shopping business. The interactive game market could also be a multibillion dollar market, bringing the total potential multimedia market to about $75 billion just on known services.

Wireless services are another field with huge

growth potential—estimated at about 25 percent a year—because current market penetration is only about 5 percent of the U.S. population. Most forecasters expect this penetration rate to grow to 20–30 percent during the next ten years. The FCC has allocated an enormous amount of additional spectrum for wireless services, to be auctioned by late 1994 or early 1995. Although the costs of acquiring this spectrum and constructing networks are still uncertain, most major cities are likely to have at least three additional companies providing wireless services within a few years. Such services will include both narrowband voice and data and, potentially, broadband services such as video and high-speed data.

To achieve their goals, companies are adopting a variety of different strategies, including the following:

■ *Move up the value chain.* As basic transport becomes more of a commodity business, companies need to move up the value chain to actual or perceived proprietary products such as owning brand names, getting more involved in the ownership of content, or developing proprietary operating systems. Thus, aside from the attraction of extending their geographic reach, the potential benefits to the telephone companies of acquiring or partnering with cable companies is two-fold: to develop expertise on the cable/programming side and to acquire ownership or participation in content. About half of the value of U S West's $2.5 billion investment in Time Warner can be imputed to the content side.

■ *Eliminate restrictive barriers.* Under the 1984 consent decree, the RBOCs are still prohibited from providing most long-distance services, a market with huge potential for them. They are also restricted, under the Cable Act of 1984, from providing cable services within their telephone service territory, although in 1993, a favorable decision for Bell Atlantic Corporation in a legal challenge to that Act will probably ultimately be applied to all telcos. Congress is actively engaged in writing communications legislation that could end both of these prohibitions. So-called CAPs—competitive access providers—which build fiber optic networks to compete with the telcos for business services are seeking to require full interconnection with the telcos and elimination of state rules that prohibit LEC competition.

■ *Eliminate rate-of-return regulation.* Multiple services are difficult and often unprofitable to provide because of the allocation methods regulatory commissions use under rate-of-return regulation. Furthermore, companies cannot fully benefit from successful marketing and cost cutting under rate-of-return regulation because most companies are at or close to their authorized returns.

■ *Enter foreign markets.* Most people believe U.S. companies are five to ten years ahead of foreign competition in their ability to operate in a competitive market. The foreign market is probably 1.5 to 2 times the size of the U.S. telecommunications market. Underdeveloped countries offer less competition and more rapid growth than developed country markets. For example, the rate of access-line growth of Telmex and the Latin American telephone companies, in a typically more favorable regulatory environment, is about 10–12 percent a year compared with 3 percent in the United States. Thus, to the extent that U.S. companies are able to invest in foreign telcos as they privatize or acquire competitive telecommunications licenses at reasonable prices, the incremental returns can be high.

■ *Get as close to customers as possible.* Every company, whether wireless, long distance, local telephone, or cable will try to get as close to their customers as possible. Doing this will involve putting together a package of attractively priced, high-quality services (telephone, long distance, wireless, entertainment), generally under a strong brand name. Ironically, this structure will be a return to the one-stop shop for telecommunication services that existed prior to the AT&T breakup. In many cases, companies will require regulatory or legislative changes, but if successful, companies will be able to spread their networks and marketing costs over a much greater revenue base and face less risk of losing customers to specialized, single-product companies.

■ *Use partnerships, affiliations, and investments to extend the company's resources and capabilities.* This strategy relates not only to company efforts to penetrate nontraditional markets (e.g., telcos getting into media) in which they do not have the core competencies but also to efforts to provide their customers with a broad range of services (e.g., MCI's investment in Nextel in order to move into wireless). Ideally, such alliances will draw on the strengths and offset the weaknesses of both parties. Cable companies, for example, have conceded that telephone and long-distance companies are much better than they are at managing networks, have much stronger financial capabilities, and are much more knowledgeable about the software needed to run networks. The cable companies are much better in programming, marketing, and so forth. Thus, cable alliances with either telcos or long-distance companies would appear to benefit both.

■ *Provide state-of-the-art services.* All companies will have to upgrade their networks to digital, switched broadband in order to provide new services and prevent competitors with their own new services from taking customers.

■ *Become a low-cost provider.* This strategy discourages new entrants and enables companies to maintain some degree of profitability even if competition erodes prices.

■ *Create strong product brands.* Companies are placing a lot of emphasis on product branding to create proprietary positions or distinctions among themselves. The most notable success thus far is MCI with Friends and Family. The RBOCs that used to sell telephone service under, for example, Ohio Bell or Illinois Bell, are using their corporate names to sell telecommunication services within their territories.

Why Strategies Differ

Most telecom companies are using only some of the strategies listed above. For example, some companies have been very active in merger/partnership activity, and others have not. The primary reason for the difference in strategies is that the corporate managements have different visions of the future and where value will lie. Clearly, the value will not be in basic transport, but no one knows if more value will lie in wireless, broadband, or operating systems. Bell Atlantic's proposed acquisition of Tele-Communications (TCI) only made sense if one believes that customers will pay for new services delivered through a broadband network and want them sooner rather than later. Other companies question that vision of the future and whether the incremental dollars from new services will accrue to the transport companies or content companies. Given that concern, aggressive acquisitions of cable companies—particularly at stock valuations twice that of the telephone companies—do not make sense.

Concerns about wireless entrants focus on the cost of acquiring spectrum and competing with the existing cellular companies. PCS is unlikely to have a meaningful cost advantage over cellular, and duplicating the national coverage that cellular has will probably take several years—if not a decade. Thus, moving into wireless through PCS carries enormous risks, although if demand is as robust as expected and a company has strong marketing and operating capabilities, the investment could have a substantial payoff.

Managements also base their strategies on their view of their companies' core competencies. Some managements view their core competency as networking and want to own transport companies—cable, wireless, or telephone. Others view their core competency as marketing and believe they can acquire enough capacity to provide either wireline or wireless services on a resale basis so they can focus on marketing and not have to own facilities. Still others believe they are better at creating operating systems and can penetrate the market that way.

Managements also have differing views of technological evolution. For most of the telephone industry's history, technology has been a plus, particularly during the past 20 years as capital costs and operating costs declined. Cost-saving technology provides huge benefits when companies lack competitors that might try to pass cost reductions through to customers. Even in a regulated environment, such technology has benefits: the company does not have to worry about applying for rate increases from regulators and can generally earn or exceed its authorized rate of return.

Technology, however, is a two-edged sword, and it has permitted competition to enter the telecommunications industry. It has been the driving factor in the trend toward cheap and powerful transmission and processing. As a result of the latter, network intelligence may move from its historically centralized function to customer premises. This trend has profound implications for the industry. If a company's capability is running a highly intelligent, centralized network, then technology could work against that company. With the capacity of fiber optics and the processing power on customer premises in the future, future networks could become simple and most of the intelligence could reside with the customer.

In essence, telecommunications networks may see the paradigm shift that the computer industry is experiencing as it moves from a mainframe to a client-server structure. The value of the network and the strategies companies adopt may change dramatically depending on the actual degree that the historical paradigm shifts. Some technologists argue that the ultimate network of the future will be a combination of robust centralized intelligence and strong processing capability on the customer side. Others argue that most of the intelligence will migrate to the end user.

Another factor in the future of telecom companies is the degree of success in entering foreign markets. Most RBOCs and some long-distance companies have tried to create foreign partnerships, and some have been successful: BellSouth Corporation has extensive cellular operations in Latin America, Pacific Telesis Group had been in Europe and Japan prior to the spin-off of its wireless domestic and international operations, and Southwestern Bell Corporation owns 10 percent of Telmex. Companies without the ability to develop partnerships to penetrate foreign markets obviously are not likely to focus on that as a primary strategy. Those that can operate in foreign markets will have enormous opportunities, particularly in the latter part of the decade, as other countries open their telecommunications networks to competition—Mexico in 1996 and the European Community in 1998, for example.

Investment Issues

The telecommunications industry's financial parameters for local-exchange, long-distance, cable, and wireless businesses are summarized in **Table 1**. In 1993, local-exchange revenues were $95 billion and long-distance revenues were $35 billion. (Reported revenues for the long-distance industry were about $60 billion, but about $25 billion of that represents access charge payments to the LECs and is excluded from their revenues on the basis that it is just a pass-through.) The cable business was about $25 billion, and the wireless business (cellular and paging) was about $13 billion.

Table 1. U.S. Telecommunications Industry: Summarized Financial Parameters

Services	1993 Revenues (billions of dollars)	5-Year Projected Growth Rate	Current EBITDA Margin
Local exchange	$ 95	3–4%	45%
Long distance	35[a]	6–8	40–45
Cable	25	7–10	45
Wireless	13	20–25	40–45
Total/average	$170	7.0%	42%

Source: Charles W. Schelke, CFA.
[a]Excludes access and other payments to U.S. telcos and foreign carriers.

In general, most analysts are forecasting the industry growth rates shown in Table 1, although this growth may not necessarily accrue to current industry participants because of new competition entering the industry. The rates shown are based on historic growth rates adjusted for the recession of the past few years. Wireless, which has been a huge growth industry since its inception ten years ago, will continue to boom at the rate of 20–25 percent a year.

The telecom industry is very profitable, as seen from the current earnings before interest, taxes, depreciation, and amortization (EBITDA) margins shown in Table 1. The EBITDA margins for all the companies are in the 40–45 percent range. Wireless is still in its early growth stage, and EBITDA margins for wireless are likely to continue to move up until additional competition enters the business.

The dominant trends in the telecommunications industry—increasing competition at all levels and increasingly rapid change in technology—have major consequences for the industry's stocks. From a telecommunications company standpoint, whether cable, local telephone, or long distance, the major issue is whether revenue trends follow cost trends, which are down, or demand trends, which are up. If they follow cost trends, telecommunications will end up in the same situation as the airline and personal computer industries. If they follow demand trends, investors will make a lot of money. Although telecom is a very profitable industry, it is unclear whether it can maintain its profitability over time given increased competition entering the industry, the technological trend toward cheap wireless and wireline transmission, and increased processing capabilities.

Four key operating issues follow from those trends: First, companies must determine what the proprietary aspects of telecommunications services will be in five or ten years. They will have to understand how to establish a proprietary position and maintain it. Second, given that the network companies generally have common technology, they will have to determine how they can differentiate themselves. Third, management must analyze the current array of strategies—entering more industry sectors, branding, moving into content, and moving into foreign markets, and so forth—to determine which are better and which they can adopt and implement at reasonable cost. Finally, the issue of whether technology trends will enhance or destroy the value of the existing network investments must be resolved.

Conclusion

In summary, the telecommunications industry is entering into a very dynamic and uncertain phase characterized by much greater competition, uncertainty, and opportunity than has been the case in the past for the major players. Given this outlook, we think investors need to be cautious in their valuation expectations on the stocks until some of the key issues noted above are resolved.

Question and Answer Session

Charles W. Schelke, CFA

Question: Will the consent decree simply be abolished to allow the RBOCs to venture into long distance, or will their entry come about through joint ventures with existing long-distance providers?

Schelke: According to the consent decree, the RBOCs cannot enter the long-distance business as long as there is any "substantial possibility" that they may use their market power to impede competition in that market. That wording has been interpreted very restrictively by Judge Harold Greene, who oversees the consent decree.

This restriction could be resolved politically two or three different ways. Ameritech Corporation and another group of five RBOCs have petitioned the FCC for a ruling that, under certain conditions, they can enter long distance. If the FCC addresses this request and lays out conditions requiring the network to be opened up, that requirement would be fulfilled in three or four years. The FCC decision would then go to Judge Green, and he would have to rule on whether the conditions of the consent decree had been fulfilled. As noted, Judge Green has been very restrictive in his interpretations, how much weight he might give an FCC ruling is unclear. He is also about 70 years old and may not preside over the consent decree in two or three years.

The more likely way the judicial restrictions could be removed or interpreted more liberally is through legislation. Congress is clearly interested in passing telecommunications legislation to remove jurisdiction from Judge Greene and to encourage development of broadband services—the so-called information highway. As of March 1994, three major bills were pending. These would permit the telcos into cable in their service territories (although prohibiting them from acquiring the existing cable company) and would move jurisdiction of the long-distance restriction to the Department of Justice and the FCC and away from Judge Greene. The proposed bills do contain the "substantial possibility" test of the consent decree. I believe this test will be met within three to four years and possibly sooner, and this seems to me to be a reasonable time period in which to expect the RBOCs to be allowed back into long distance.

This three- to four-year time frame is based on the prospects that the RBOCs will have extensive network competition at the end of that time period. Competitive interconnection, for example, is becoming a reality. The FCC has mandated interconnection by competitors for both switched and private-line services, and state regulators are following suit. Resolution of the details—cost allocation and how much can be charged for interconnection—will be a major step forward in opening up the RBOCs' networks.

A second step toward providing competitive local-exchange services will come through PCS. Licenses are expected to be auctioned by late 1994 or early 1995, and the networks built within two or three years. These systems will present a viable alternative to the wire network for local telecommunication services. Thus, within three or four years, the RBOC network will be opened up.

As far as how the RBOCs enter the market from a business standpoint, I expect them to focus on their in-region markets. Most of the companies already connect their major exchanges by fiber optic transmission for their own internal traffic, so they could easily upgrade that capacity to carry customer traffic.

Not only is the in-region market easy for the RBOCs to enter but also it is extremely attractive. For most of the RBOCs, roughly half of the calls that originate in their service territories also terminate in their service territories. That is a lot of traffic that they can handle completely within their territories. About 60 percent of the long-distance industry's revenues come from ordinary long-distance services—residential services and small to medium-sized business services—so the RBOCs do not need to install the sophisticated software needed to service the Fortune 1000 customer base. The RBOCs could move into the long-distance market on a very low-cost basis because their incremental costs of serving this market would be very low. They already have the networks and the billing systems, because they bill for the long-distance companies, and they already market to the customer base—the residential and small-business customers in their service territories. Long distance would be a big plus for them and, by extension, a big negative for the long-distance companies.

Question: How can an investor determine the extent to which a firm is adapting to this transition period? You seem to imply that all of the RBOCs are trying to provide the fullest array of services. Please describe who is doing what and whether a guiding or preferred strategy exists.

Schelke: The best strategy is not clear. The choice of strategy partly depends on price—how much does a company have to pay for acquisitions or partnerships to expand its geographic or operating capabilities or position? For example, Bell Atlantic proposed to acquire TCI at a cost of 11.75 times cash flow when Bell Atlantic stock was selling at 6 times cash flow. Given TCI's size and the consequent dilution to Bell Atlantic, this did not seem to be an attractive acquisition from a Bell Atlantic shareholder standpoint. If the vision of multimedia works out as the bulls expect, however, and it works out sooner rather than later, the acquisition could be justified. This example points up the difficulty of prejudging strategies. Nevertheless, I do like certain strategies:

- Entrance into foreign markets, again at reasonable prices.
- Exposure to cable and media to a moderate extent, but not on a huge, highly dilutive basis.
- Expansion in-region, assuming that regulation is favorable or trending in a positive direction.

In reality as investors, we have to look at each company action on a case-by-case basis. For that reason, I like to value the stocks on a sum-of-the-parts basis, and because the future is so uncertain, I don't like to pay up for perspective synergies or developments.

Despite some reservations about wireless and the cost of entry into PCS, in general, I think a wireless strategy is strong, although with four or five wireless companies, price competition may become intense. For an industry growing as rapidly as the wireless industry, that degree of competition may not create much of a pricing problem until the industry reaches a stage of maturity at which 30–40 percent or more of the U.S. population are wireless customers.

Question: What impact will foreign telcos have in the U.S.?

Schelke: Two foreign companies are in the U.S. market: Cable and Wireless (a relatively small long-distance operation) and the British Telecom–MCI partnership. AT&T World Source is trying to develop partnerships with foreign countries. That effort is directed at serving the large business user market, the multinational Fortune 1000 companies that want transparent services. I would be surprised to see foreign telephone companies entering the U.S. market, because it is already so competitive. Thus, the British Telecom–MCI agreement may be somewhat unique in terms of foreign telephone companies investing in U.S. properties. More likely will be the reverse—which is already occurring—of U.S. telecom companies investing outside the United States.

Question: What countries provide the best opportunities for RBOC international expansion?

Schelke: Third World and lesser developed countries offer tremendous opportunities because their GDP growth rates are higher than those of developed countries, they are "under telephoned," and regulation is generally favorable. The Latin American market, for example, is very attractive. A number of Asian countries will privatize or have already privatized their telecommunication systems, and these will offer investment opportunities. Most of Latin America and all of the developing countries of Asia and South Korea are growing at two or three times the rate of the United States, and telephone penetration is one line for every eight or ten people as opposed to one line for every two people in the United States. Many of these countries also have stable governments and political systems. Thus, to the extent that U.S. companies can enter those markets on reasonable terms, it could be a big plus for them.

Question: Are foreign markets poised to bypass wire technologies and perhaps move right into wireless cellular PCS?

Schelke: Yes. The logical thing to do in countries with undeveloped wire-based telephone systems would be to set up a cellular/satellite system and, over time, lay down a fiber optic or fiber/coax system instead of using copper (twisted pair) technology. To a certain extent, however, many of these countries are still building with traditional technology, although they are also installing wireless technology at the same time. The economics clearly dictate using wireless in order to dramatically expand telecommunications capabilities in a relatively short time frame. Also, this system would not be made obsolete by a wired infrastructure because the wireless system can always be used as a secondary system purely for mobility purposes. What seems to be evolving is a mix of wireless and wireline, the latter a mix of new and older technologies, but in general, much greater reliance on wireless than is required in countries with effective wireline systems.

Question: What business risks do the foreign expansions of RBOCs pose? Will the international exposure affect their credit ratings?

Schelke: I am not a credit analyst, and I do not focus too much on that aspect of the industry. Typically, entrance into a foreign country does not cost that much. Companies usually do it as minority partners because of country restrictions. For $300 million to $500 million, a company can generally enter most countries as a partner and construct a fairly robust telecommunication structure. Thus, for companies the size of the Bells, with capital bases of $20 billion to $30 billion, investments of $1 billion to $2 billion—even if all of it is borrowed—do materially impact their credit positions, particularly given the potential returns from those investments.

In terms of business risks, you have to look at the individual investment. In general, however, we think the business risks (using the term in a narrow sense) are low because of the huge demand in many countries for expanded telecommunications services and government interest and recognition that economic development requires a modern telecommunications infrastructure. As with any telecom company, investors need to understand the regulatory arrangements and the time frame in which the company will have an exclusive franchise. In a broader sense, the major business risks that we see in investing in foreign telecom companies are related to the country itself: political stability or instability, currency risk, and the inflation and growth outlook.

Question: Does your outlook differ from the consensus for the future of the telecommunications industry?

Schelke: I am probably more positive than other analysts on the outlook for the telephone companies and less positive on the long-distance companies. Many investors have been very negative on the telephone industry for the past few years because competition is entering the local-exchange segment. Because of regulatory and other restrictions, the fear is that these companies are sitting ducks for competitors. I believe there will be some positive offsets, primarily reduced regulation as competition increases and the potential to move into new markets such as cable and long distance. In regard to the latter, I believe that a local-exchange company can more easily move into long distance than a long-distance company can move into local exchange, because the long-distance network is essentially in place for the RBOC. Thus, I am more concerned about the competitive issues for the long-distance companies going forward, so I tend to assign a lower valuation to the long-distance companies than other analysts may. Like most analysts, I am optimistic about the outlook for wireless. The only question is how competitive that business becomes and how many wireless companies will enter the market after the PCS auction.

Question: In your valuations, under what circumstances would you use EBITDA instead of net earnings, and what are your expectations for normalized post-transition margins?

Schelke: The companies must work very hard to maintain margins at current levels. They may pick up something on the revenue side, but current margins are very healthy, and I do not believe the industry is capable of sustaining those as more competition enters the market toward the end of the decade.

My valuation approach tends to be fairly eclectic. We break down the telephone companies into their business components—local-exchange service, cellular operations, and other businesses such as international ownership—and try to value them separately. For these separate valuations, we use either relative price–earnings ratios (P/Es) or cash-flow multiples, generally the former for telephone operations and the latter for cellular and other businesses. We also relate those values to the growth rates we foresee on a business-as-usual basis—that is, without the cable or long-distance increments.

We estimate LECs' earnings per share should grow 5–9 percent, with 3–5 percent of that coming from the telephone side and 2–4 percent from the cellular side; for companies that have some international exposure or other businesses, maybe 1–2 percent will come from that. On a P/E valuation basis, I believe these stocks should trade anywhere from a 10 percent discount to about a 10 percent premium to the market, based on this 5–9 percent EPS growth outlook. This would also equate to about six to seven times cash flow on overall corporate numbers for the telcos.

For the long-distance companies, I basically use relative P/E multiple valuations. Although the current growth being reported by most long-distance companies would justify significant premiums to the market for many stocks, my concern about increased competition from the Bells is such that I generally assign only a market multiple or a 10 percent premium to the market as a valuation target.

Question: What effect will the expected increase in wireless competition have on margins?

Schelke: Wireless will still be a duopoly for the next two or three years, and EBITDA margins for wireless companies will probably move up to the 45–55 percent range or perhaps even higher during that period. The impact of PCS on cellular margins during the 1990s is not clear. Initially, PCS will undoubtably have to price-discount relative to cellular, because for the first few years, PCS services will not be as robust as the existing wireless services. By the time PCS takes hold, cellular companies will have had at least 14 years to build their networks. Cellular will offer seamless service throughout the United States, and customers will be able to travel from one city to the next with roaming that is transparent to the user. In contrast, for its first few years, PCS will be more of a neighborhood or city service than a regional or national service. PCS used in homes or neighborhoods as a wireless telephone service will probably be offered at flat rates rather than on a per minute basis. Use within five or ten miles outside of the neighborhood might be charged at 5–10 cents a minute and for intracity or regional use, PCS will probably be priced close to cellular service—20–25 cents a minute. Cellular companies will probably respond to PCS competition by adopting their pricing but offering slightly better services—larger service areas and greater roaming ability—and focusing on the customers that require higher quality services. If the cellular companies are adept, margins could remain relatively high for several years after PCS services are introduced. Ultimately, however, with four to five wireless companies, margins will undoubtably come down, but this may be more of an issue very late in this decade or early in the 21st century than during the next three to five years.

Question: Will the telecommunications company of the future need a critical size or critical mass?

Schelke: In terms of scale economies, RBOCs are about as large as they need to be, in my opinion. The trend in telecommunications equipment is toward declining costs, so starting a telecommunications company is becoming cheaper. I see no significant economic cost benefits from aggregations of telephone companies larger than existing telephone companies or long-distance companies larger than existing long-distance companies. There may be some modest scale advantages in nationwide marketing, but the current size of the companies certainly seems sufficient to provide any economies that are available.

Question: How do satellite communications fit into the transition period?

Schelke: Satellite use will become increasingly specialized. Currently, about half of international traffic is carried over satellites. From a cost standpoint, satellites are ideally suited for long routes with low-density traffic; relative to fiber optic cable, they are not economically attractive for high-density traffic or relatively short routes such as the North Atlantic. Currently, satellites over the North Atlantic are used primarily as backup for cable facilities, but they are used extensively as primary facilities for the low-density traffic on the long routes to Asia and South America, and they will continue to play a major role in basic transmission there.

The other major role for satellites is in the provision of mobile communications in applications (marine, aeronautical, low-density terrestrial) for which land-based wireline or radio services are not available. Cost has a significant impact on the use of satellites for mobile communications. Currently, mobile services via satellite cost about $6–$8 a minute compared with $0.25–$0.35 for cellular. Even new satellite systems or technologies are only likely to cut the cost to $2–$3 a minute. With those high costs, satellites will only be used for mobile communications in areas that do not have access to land-based radio or landline service. Even as specialized carriers, I believe that satellite use will grow at a rate of perhaps 10–20 percent a year depending on the particular application.

Question: Do you have any thoughts on the status of current and anticipated legislation?

Schelke: We are likely to see more legislative activity in the next two or three years than we have since 1934, when the last major act affecting this industry, the Communications Act, was passed. That act is clearly outdated by technological and competitive developments in the industry. If Congress examines the cable act, it may open up the whole issue of a rewrite of the telecommunications act. One way or the other, the industry will have some legislation from Congress. It may be very narrow and only apply to cable, or it may be much broader than that and cover such issues as the terms and conditions under which RBOCs can provide long-distance service. As of March 1994, the broader approach seems much more likely.

Internal and External Factors Affecting the Local and Long-Distance Telecommunications Industries

Stephanie Georges Comfort
Principal
Morgan Stanley & Company, Inc.

> The telecommunications industry is driven by a number of interrelating influences, and it is undergoing some far-reaching technological and structural changes. Both of these circumstances make this industry particularly difficult to forecast.

The telecommunications industry is in the midst of dramatic and rapid change. One need look no farther than the popular press to discover that the telecommunications services industry—landline telephone, long-distance, and wireless communications—is rife with change.

External Factors

Certain parameters are endemic to the industry and are part of each company's decision-making process. These external factors can be divided into five categories: regulation, competition, economic cycles, technological advances, and political stability. Both long-distance and local companies are redefining their strategies in response to developments in these five important areas.

Regulation

Regulation occurs on four levels: federal regulation (through the FCC), state regulation (local public utility commissions and public service commissions), the judiciary (Judge Harold Greene, who handed down the initial consent decree responsible for the breakup of AT&T; administrative law judges; and district courts), and the lawmakers (the U.S. Senate and House subcommittees).

Changes have been dramatic at the FCC. Under Chairman Dennis Patrick, the commission implemented price-cap regulation for AT&T (price caps allow price-based rather than return-based regulation), which set the wheels in motion for many state authorities to implement similar structures in their jurisdictions. Following those changes, the FCC moved to open up the interconnect business—first special, then switched—by allowing competitors to "co-locate," that is, have access to the local phone company's central office.

The FCC has also taken the initiative in introducing competition in the wireless business by allowing specialized mobile radio (SMR) companies to use their spectrum for two-way digital mobile radio service and by defining PCS auctioning process expected to begin in May 1994. Finally, the FCC proposed a video dial tone (the ability to carry video signals over the telephone network) ruling that would allow the Bell regional holding companies (RHCs) to use their networks to provide video dial tone services but would continue to bar them from providing programming.

Regulatory policies and changes vary in individual states; some state regulators have moved faster and been more flexible than others. The plans implemented have ranged from the maintenance of strict rate-of-return regulation (New York) to sharing plans, which allow the phone company to retain a portion of the earnings above a threshold return (New Jersey and California), and price caps (Florida and Illinois). Other states have opened the local bottlenecks by allowing competitors to provide local dial tone; that is, customers in these states may have a complete alternative to the local telephone company. For example, Metropolitan Fiber Systems' (MFS) Intelenet provides customers with a local dial tone alternative to New York Telephone.

On the judicial front, the district court in Alexandria, Virginia, surprised Wall Street with its August

1993 ruling that declared the 1984 Cable Act unconstitutional. The Cable Act prevented a Bell RHC from selling video programming to subscribers in its own telephone service area. The ruling has been appealed, but it set the wheels of change in motion. Moreover, it could advance congressional efforts to restructure the telecommunications industry through the implementation of new telecom policy.

As for the lawmakers, a House subcommittee and the Senate Commerce Subcommittee on Communications have held hearings to help develop telecom policy. The primary proposal to date is the Telecommunications Infrastructure Act of 1993, which would allow the Bell RHCs to provide some cellular, cable television services, and possibly long distance across local access transport area (LATA) boundaries and to provide cable television offerings in their respective operating regions.[1] During the next several months, both the House and Senate will advance policy on the opening of local exchanges to competition, requiring phone companies to provide interconnection to competitors, and imposing universal service requirements and safeguards to eliminate subsidies.

Competition

Related to regulatory changes and equally dramatic are the competitive forces that are defining the future of the industry. Competition is becoming pervasive. The landline network is being invaded by cable television companies (or former RBOCs in disguise) that plan to use their infrastructure to compete with the established local phone companies.

The wired network is threatened by the improving capabilities and increasing penetration of the wireless network. The cellular industry is expected to reach 6 percent penetration in 1993; 50 percent of all new wired and wireless lines placed in service by that time will be cellular phones. Moreover, personal communications networks are just around the bend. The FCC has begun the process of opening 200 megahertz of spectrum for PCS, which could ultimately compete with the landline network.

The long-distance industry has demonstrated, however, that competition does not have to be disastrous. The segment has experienced heavy competition. The result has been reduced long-distance prices (by 55 percent during the past decade) and increased use, which is reflected in compounded annual volume growth of an average 7–8 percent. Furthermore, the evolution of competition in the long-distance business may not be over. The rumblings that the Bell RHCs will provide some form of long distance are growing louder, and some early signs, including provisions in telecom policy bills being introduced in Congress, point to this development. Although the timing of such changes is not yet foreseeable, the local telephone companies could ultimately be allowed in long distance as a quid pro quo for a competitive local loop.

Competition is coming from new entrants, such as the CAPs, which originally took advantage of opportunities to bypass the local-access business. These companies are migrating toward the provision of other local telephone services on a state-by-state basis as regulation permits. On the wireless side, the SMRs, with the help of Motorola, have developed a technology that could position them to compete for the wireless communications business.

Economic Cycles

Another external influence is the economy; some of the relationships are obvious.

- *Inflation.* Inflation affects the cost of capital, as well as interest rates. It thus affects network buildout, required returns on investment, and the pricing of services as determined by regulators.
- *GDP growth.* Growth in the economy is correlated with the growth in access lines for local telephone companies and with the growth rate in minutes of use (or volumes) for the long-distance carriers. This is shown in **Figure 1** and **Figure 2**.
- *Residential building.* Figures for housing starts may be better predictors of industry growth than GDP figures. Housing starts reveal logical trends in the growth of residential access lines and, therefore, the growth outlook for local telephone companies' revenues. These measurements can add particular insight into trends by regions.

Technological Advances

Technology and changes in technology continue to have an impact on both internal and external factors affecting the telecommunications industry. As an external factor, advances in supplier and adjunct technology force changes in the telecom services companies. For instance, the declining cost of fiber optics has allowed telephone companies to install fiber rather than copper wire in all new builds. The miniaturization in size (and price) of cellular handsets has changed the mobile telephone business into a wireless PCS business. It has also propelled mobile penetration and use. Signaling system seven (SS-7) capabilities in the telephone network allow the introduction of services such as voice mail, voice messaging, and caller identification.[2] Such services

---

[1] A LATA is one of 161 local telephone exchange areas established as a result of the AT&T divestiture. LATAs serve to distinguish local phone service from long-distance service.

[2] SS-7 is a specific network control system that, at the originating end of a call, uses a packet switch to send a query to a data base. SS-7 accommodates enhanced 800 service, wide-area centrex services, and other types of advanced telecom services.

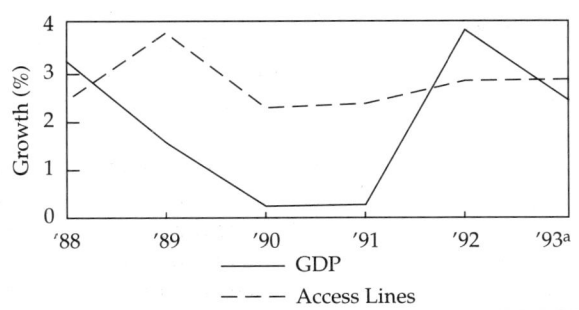

Figure 1. Growth in Real GDP versus Growth in Access Lines in Service

Source: Stephanie G. Comfort based on company reports and Morgan Stanley Research.
aEstimate.

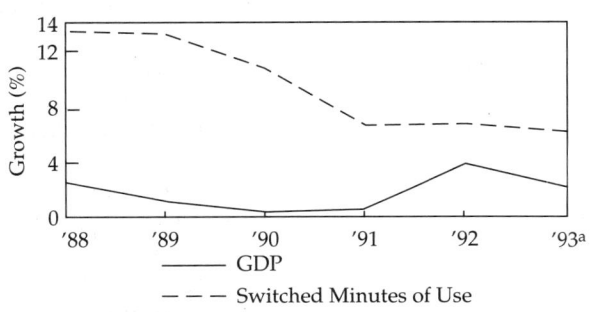

Figure 2. Growth in Real GDP versus Growth in Switched Minutes of Use

Source: Stephanie G. Comfort based on company reports and Morgan Stanley Research.
aEstimate.

are higher margin, value-added services than local phone companies' traditional offerings to customers.

Other new technologies include asynchronous digital subscriber line, a compression technology that allows twisted pairs of copper lines to carry video signals, and asynchronous transfer mode (ATM), a technology that allows data to be transferred economically, rapidly, and in large quantities across the local and long-distance network. ATM is typically described as a "killer" technology, one that is likely to revolutionize the way we use and move data.

All of these technological advances change the way in which to evaluate the network of local and long-distance services: the robustness of the network (i.e., its capabilities), its efficiencies, and its ability to add value for consumers.

Political Stability

Finally, the globalization of many of the telephone companies raises issues involving the stability of foreign countries. One factor that company managers must assess is government control, or broadly defined, the role of the government in the telecom industry and governmental influence on regulations in the industry. For example, part of the attraction of investing in Telmex (the telecom company of Mexico) for Southwestern Bell Corporation was linked to the Mexican government's clearly stated interest in developing the telecommunications infrastructure in Mexico and its willingness to structure a flexible and attractive regulatory system. Other globalization issues include currency and the ability to repatriate capital.

Internal Factors

Companies providing local and long-distance services are affected by many internal industry factors, but six are basic: management, labor relations and costs, the state of the network, marketing strategies, diversification efforts, and financing capacity.

Management

Management is often ignored in financial analysis, but it is a significant factor when differentiating one company from another. With the changes occurring in the industry, management style has become especially important. Management style can be divided into three categories: strategic approach, composition, and structure.

Vision and bold strategy are the result of a proactive rather than a reactive approach to the business and to opportunities. Proactive policies have led to such partnerships as U S West and Time Warner, Southwestern Bell and Metromedia Paging, and the proposed union of AT&T and McCaw Cellular Communications. These strategies may be risky, but without taking the risk, a company in this industry could end up without any meaningful growth opportunities—or have to pay too much for them.

The composition of management is particularly important to consider in the local and long-distance phone sector because many of the managers have long histories of working for the old, monopolistic Bell and their approach to the business is steeped in that environment. A number of companies, however, have been bringing in managers from outside the industry to help reposition the company and its thinking. U S West, MCI Communications Corporation, and AT&T have been rethinking the composition of management.

Finally, corporate structure can be an issue. Whether the structure is bureaucratic or entrepreneurial can have implications for a company's flexibility to deal with industry changes. Talking to managers and gaining an understanding of a company's systems will usually reveal the type of structure. Although the Bell RHCs are far from being entrepreneurial entities, several have restructured and reduced their bureaucratic layers.

Labor Relations and Costs

A second significant internal factor is cost structure—in relation to overall operating costs and labor costs. The two most common measures of efficiency used in comparing the Bell RHCs are cost per access line and cost per employee. As shown in **Table 1**, costs per access line have declined for all of the companies during the past five years; the more dramatic declines have occurred in the past two years. Ameritech Corporation, specifically Illinois Bell (one of its subsidiary telephone companies), and U S West have aggressively worked costs down; NYNEX Corporation and BellSouth Corporation still have some work to do.

Table 1. Cost per Access Line
(thousands of dollars)

Company	1988	1989	1990	1991	1992
Ameritech[a]	$405	$417	$408	$413	$396
Bell Atlantic[b]	473	489	491	477	476
BellSouth	553	555	563	556	549
NYNEX	603	613	595	605	590
PacTel	545	523	515	523	498
Southwestern Bell	479	477	472	460	472
U S West	522	518	496	488	487
Average	511	513	506	503	495
Year-to-year growth	—	0.3%	-1.5%	-0.5%	-1.6%

Source: Stephanie G. Comfort based on company reports and Morgan Stanley Research.
[a]Based on Illinois Bell.
[b]Based on New Jersey Bell.

Similarly, the number of telephone employees per 10,000 access lines has declined from an average of 51 in 1988 to an average of 40 as of year-end 1992 **(Table 2)**. Ameritech, Bell Atlantic Corporation (New Jersey RHC), and U S West stand out as aggressive on employee costs, while NYNEX and BellSouth are at the high end of the group.

State of the Network

The third important internal factor when analyzing the local and long-distance telecom sector is the state of the network. The focus in this discussion will be fiber deployment and digital switching, although network capabilities cannot be fully captured by only two sets of statistics.

Digital switching capability expanded from 1987 to 1992. As seen in **Table 3**, some RHCs had as few as 17 percent of their lines served by digital switch in 1987. Now, for the most part, these companies serve 50 percent or more lines digitally.

Table 2. Telephone Employees per 10,000 Lines

Company	1988	1989	1990	1991	1992
Ameritech[a]	46.5	44.9	42.7	40.0	37.5
Bell Atlantic[b]	44.7	41.9	41.3	38.6	34.7
BellSouth	54.1	51.1	49.1	45.5	44.4
NYNEX	57.8	55.2	51.9	46.8	45.2
PacTel	51.4	48.5	44.5	41.0	38.8
Southwestern Bell	51.1	49.6	48.1	42.2	39.0
U S West	48.4	47.6	43.7	42.5	39.3
Average	50.6	48.4	45.9	42.4	39.8

Source: Stephanie G. Comfort based on company reports and Morgan Stanley Research.
[a]Based on Illinois Bell.
[b]Based on New Jersey Bell.

Table 3. Network Access Lines Served by Digital Switch Offices, 1987–92

Company	1987	1988	1989	1990	1991	1992
Ameritech[a]	24%	30%	37%	40%	46%	53%
Bell Atlantic[a]	28	34	41	49	56	63
BellSouth	30	38	44	51	57	61
GTE[b]	57	66	70	74	77	82
NYNEX	24	38	49	55	60	67
PacTel[b]	19	23	27	35	40	43
Southwestern Bell	18	21	23	26	32	39
U S West	17	24	29	35	41	48

Sources: Stephanie G. Comfort based on company reports and Morgan Stanley Research.
[a]Reported as "percentage of digital lines" for entire corporation.
[b]Reported as "percentage of local end office switches."

In the category of fiber deployment, discrepancies among companies are more pronounced. As seen in **Table 4**, Bell Atlantic, which has made clear its strategy to compete in video and information transmission both in and out of its region, has invested in more than a million fiber route miles. In contrast, Pacific Telesis Group (PacTel) has focused its strategy on deploying fiber in major urban areas on an as-needed basis and has deployed about 300,000 fiber route miles.

Marketing Strategies

Because the industry has historically operated in a monopoly environment, marketing has not been at the forefront of management initiatives. For a long time, the name of the game has been market share. Pricing has represented the vehicle for attracting customers and building share. As shown in **Figure 3**, the pricing trend has been downward and the price per switched minute of use declined from 25 cents to 16–17 cents between 1985 and 1992. At the same time, volume growth helped propel revenues; thus, revenue growth remained strong through the difficult pricing environment.

The focus of marketing in the long-distance game, and to some extent in the local telephone and

Table 4. Fiber Deployed, 1987–92
(thousands of route miles)

Company	1987	1988	1989	1990	1991	1992
Ameritech	148	180	230	307	401	586
Bell Atlantic	187	274	373	523	765	1,163
BellSouth	218	319	445	609	769	939
GTE	NA	190	244	317	360	NA
NYNEX	207	291	358	473	637	807
PacTel	102	108	128	185	246	304
Southwestern Bell	169	215	270	372	478	576
U S West	96	137	235	352	542	797

Sources: "Fiber Deployment Update," FCC (March 1992).
NA = not available.

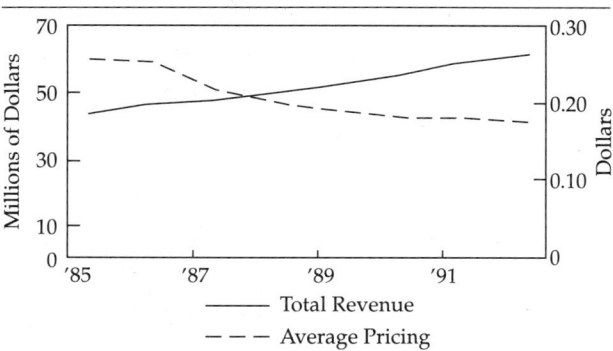

Figure 3. Pricing per Interstate Switched Minutes of Use

——— Total Revenue
— — — Average Pricing

Source: Stephanie G. Comfort based on company reports, FCC, and Morgan Stanley Research estimates.

wireless businesses, is now moving from pricing to branding and brand loyalty. Brand names that have become familiar include Mobilink, Friends and Family, AT&T, Sprint, and Cellular One. The result is a stable price environment and product differentiation. As new entrants approach the market, marketing strategies are expected to become increasingly important in differentiating the Bell RHCs—and in defining their relative success in the industry.

Diversification Efforts

Originally, diversification for the Bell RHCs meant adding assets in other, unrelated businesses—such as real estate, software, financial services, and computer stores—in order to offset the slow growth of the telephone business. Now, many telecom companies, recognizing that the growth opportunities in their core communications and related businesses are more attractive and suitable than other businesses, have divested those nonstrategic assets. NYNEX, for example, has sold its hardware and software units, and both U S West and Bell Atlantic are exiting the financial services business.

Telecom managers can characterize their diversification strategies in three ways: in-region versus out-of-region, international versus domestic, and vertical versus horizontal diversification. The Bell RHCs can be divided according to diversification strategy as follows: Ameritech and NYNEX have taken a "stick to the knitting" approach and concentrate within their own core operating regions; Bell Atlantic, BellSouth, PacTel, Southwestern Bell, and U S West have ventured out of their core operating regions. With purchases of Paradyne and Italtel in equipment manufacturing, NCR and Teradata in computers, four software companies, four interactive companies, and Eaton Financial and Universal Card Services, AT&T has made a clear move toward full vertical integration—from long distance to wireless to equipment.

Financial Capability

The final internal factor is the financing capabilities of the telecom companies. A strong balance sheet is important in this industry to take advantage of the opportunities for diversification and consolidation and to make the large capital investments envisioned. Two revealing measures of financial wherewithal and indications of constraints are debt ratios and debt ratings.

The average leverage and debt ratings vary among the four key telecom sectors: local, long distance, cellular, and cable. As seen in **Table 5**, the local telephone companies appear to have the greatest degree of financial latitude; the cable operators and cellular companies are heavily leveraged. These distinctions help explain some of the transactions and proposed transactions such as the proposed merger between AT&T and McCaw and between Tele-Communications (TCI) and Bell Atlantic.

Table 5. Financing Capacity

	Debt Ratio	Debt Rating
Local telcos[a]	46%	A+–AA–
Long distance[b]	53	A
Cellular[c]	77	BB–BBB
Cable[d]	85	BB

Source: Stephanie G. Comfort based on company reports and Morgan Stanley Research estimates.

[a]Includes Ameritech, Bell Atlantic, BellSouth, GTE, NYNEX, PacTel, Southwestern Bell, and U S West.

[b]Includes AT&T, MCI, and Sprint.

[c]Includes Cellular Communications, Centennial Cellular, Contel Cellular, LIN Broadcasting Corporation, and Vanguard Cellular Systems.

[d]Includes ComCast Corporation, TCA Cable TV, TCI, and Viacom Cable.

Conclusions

The influences of factors, internal and external, on any sector in telecommunications will result from the interrelationships among the factors. Moreover, the factors and the relationships are dynamic. Diversification, for instance, will depend on competitiveness, and regulations could constrain or encourage it. Technological advances will influence costs and network efficiencies. Management approaches will have an impact on each company's competitive positioning. All of this complexity makes forecasting the industry and its direction a task that is bound to have some margin of error, but it also makes following this sector lively and challenging.

Question and Answer Session

Stephanie Georges Comfort

Question: How competitive do you expect this industry to be, given the culture in which it is steeped? To what extent might the proactive management of one firm lead to reactive management strategies on an industry basis?

Comfort: Morgan Stanley believes the industry can be very competitive, which raises a great many issues for the historically monopolistic Bell RHCs but does not mean the stocks are not attractive. The future could include three or four ways to provide services into the home, with some niche providers and some full-service providers. Based on the consolidations, the future will contain at least two hard wires to the home. In addition, the duopoly clearly establishes two wireless providers, and then a third potential competitor, SMR, has been building steam and has experienced great acceptance in the stock market. The fourth, fifth, sixth, seventh, and eighth competitors are potentially PCS providers. This business is certainly not going to be a duopoly.

One of the industries in which Morgan Stanley looks for benchmark companies, because of the many comparisons with the telecom industry, is network broadcasting by such companies as CBS or Capital Cities/ABC. The industry is fairly heavily regulated but competitive, transitions have been very difficult, and margins have been declining. The average valuation of the network companies has been based much more on cash flow than earnings and involves, at best, an 8 or 9 multiple, depending on the company.

In local and long-distance telecom also, margins will probably decline. At the same time, opportunities for new businesses will arise. Thus, based on the experience of the network industry, with the current valuations of the Bells at 5 or 6 times cash flows, some room for upward valuation will probably exist among the Bells even in a competitive environment. Competition has not been bad for the consumer or the industry and will not be all that bad for the stocks.

Some nuances certainly must be considered in the issue of proactive versus reactive management. As one company becomes proactive, another must become reactive, and the roles may eventually reverse. The Bells seem to be in a box, however, and must be proactive if they are to get out. A Bell RHC that does not diversify out of its region and does not have proactive management, either structurally or operationally, will be operating as if it were a monopoly in an industry that will be very competitive.

Question: What are the differences between the RHCs and the independent companies in this transitional environment and in the environment after transition?

Comfort: The Bells and independents are one continuous group of local-exchange companies that have bottleneck or monopoly positions. All wireless and wireline independents face a more competitive environment than in the past. Thus, the independents have challenges similar to those of the RHCs, although they do not, traditionally or under the modified final judgment, have as heavy regulatory restrictions. The independents have had some flexibility to take the same risks that the Bells have been taking, but many have steered away from those strategies from lack of either size or interest.

Challenges to the independents may not be in the same time frame or of the same degree as in the RHCs' case. Some of the rural companies, for example, may not experience competition as early as the RHCs, but they and other companies will ultimately take advantage of pricing discontinuities.

Question: What effects will the expected market expansion have on the competition for capital, the RHCs' capital structures, and the trends in costs of capital?

Comfort: That effect is the crux of our valuations. Capital expenditures may not be increasing, but they are continuing at a high level. In addition, because of the many consolidations and the prices paid, access to the capital markets will increase. Based on offerings publicly known or being considered, the domestic industry alone, excluding international operations, will raise more than $10 billion in equity during the next two years.

To keep the rating agencies happy and to keep access to the capital markets fairly safe, the local and long-distance telephone companies will use a mixture of debt and equity to finance some of the build-out.

Question: When you evaluate capital-structure issues, do you focus on market versus book values? How should investors measure an appropriate debt capacity or capital structure?

Comfort: Morgan Stanley includes traditional measures of capital structure—that is, the ratio of market to book value—but such measures have become less meaningful nowadays because they put too much constraint on some companies. Because the industry is in a transition period, the companies must be given some latitude. Generally, we look at price–book ratios, but we have been moving toward the ratio of price to cash flow as a valuation measure in order to give the companies more leeway. We have also used the ratio of debt to total capital as a measure when a company is approaching the danger zone (e.g., debt ratios in the high 50 percent range), particularly with the rating agencies and debt investors.

The market reaction to industry moves has been interesting. If I had been asked a year ago, when it was easier to be an analyst, what would happen to the stock of some unnamed Bell company if it made an acquisition that was 25 or 30 percent dilutive to earnings, I would have said: Reduce the earnings by 30 percent and trim the stock price by a commensurate amount. Well, following the announcement of the proposed merger of Bell Atlantic and TCI, the company stock jumped 8 points. The market is ahead of the analysts in assessing some of this transition, is willing to accept highly dilutive acquisitions as long as they are strategically right, and is willing to stretch the definition of debt capacity and equity raising to include funding in general.

Question: Which specific costs are the focus of measuring efficiency, and what is your interpretation of the figures for costs per access line? How important is the particular service territory of the RHC when evaluating costs of servicing?

Comfort: The cost measures given in Tables 1 and 2 are only two of the measures we examine in terms of efficiency. Companies have differences in cost structures. Some companies (such as U S West) are rurally based, and they have much higher costs than some urban companies; some companies (such as GTE Corporation) are less centralized than others; and some companies (such as NYNEX) operate in very high-cost environments. So, comparisons are not clear and simple.

Technology is upgrading the networks, however, and making them more efficient than in the past, so cost structures will become more similar and comparable among the RHCs. Once the industry is beyond this phase of aggressive investing in the networks and into a phase of maintaining networks remotely through digital technology or fiber optics, comparisons will be more meaningful. Even now, despite the differences among territories, operating margins have typically clustered; the pace of improvement in margins has varied, but it has generally been in the same direction for all the RHCs. Therefore, figures for cost per access line and employees per 10,000 lines converge. Nevertheless, today's analysts must consider the differences in cost structures, the regulatory environment, and the difficulties in serving rural areas and adjust the valuations accordingly.

Question: Please elaborate on the investment implications of the reduction in work force for the RHCs. To what extent do you think the state regulatory commissions are looking at union-related issues when they regulate the local service companies?

Comfort: As the companies put in more effective and efficient systems that can do more and more of the work remotely, the need for labor will be reduced. In the interim, while the technologies are being implemented, consumers—business and residential—are experiencing a service gap. For example, the last time I moved, it took several tries to get someone to change my phone. Efficiency of service had declined since my previous interaction with the local telephone company. So the reduced work force is having a pronounced effect on the quality of service. This gap is something the companies must manage.

Technology will help. For example, U S West recently said it was aggressively (beyond what it has already done) reducing the number of its service centers and people, but at the same time, the company is implementing software that makes that transition transparent to the consumer.

Work-force reductions are a difficult problem, and regulators are struggling with the issue. I think state regulators are aware of the pros and cons. The governors, local regulators, and all elected and appointed officials are concerned about their local economies and unemployment; they are concerned that these companies are reducing their work forces. But they also recognize that competition is coming. Forcing the Bell companies to maintain high cost structures would really not be fair in this environment.

Question: Will the RHCs have to write off their investments in copper wire? What is the remaining useful life of the existing telephone networks, and do depreciation charges reflect this factor?

Comfort: In the transition period, the issue of copper obsolescence will present a very difficult problem for the industry. The local telephone companies in particular must wrestle with the issue. AT&T handled it by simply taking a $1 billion write-off each year for several years.

In general, the problem the Bell companies are facing is that they can do all they want to accelerate depreciation and become more competitive, but the regulators, particularly for a company under strict rate-of-return regulations, make the final judgments about the companies' abilities to reflect costs in pricing.

The good news is that regulators are considering some of these issues and are becoming more permissive. The FCC has already taken steps toward accelerating the depreciation of the copper network.

U S West has announced that it will accelerate the depreciation on its financial books. That move does not mean much in terms of pricing, but it did affect the stock, which increased 2 points. What it really meant is that U S West is going to focus on this issue. In effect, U S West was saying that the issue is important, and it is going to force the regulators' hands. Such actions will be the stimulus for change. The companies will take it upon themselves to make the regulators aware of the competitive environment and will get the FCC, for example, to move more aggressively to implement change.

Question: What will happen to the demand for copper in light of the technological changes?

Comfort: At the moment, copper is not dead, but companies prefer fiber optics for new builds. As companies indicate they want to go the route of video or interactive multimedia, more and more of the new investment will be in fiber, and copper will eventually phase out. Thus, investors in copper need to recognize that most of these companies have either made or are planning to make the transition to fiber optics. For example, U S West has indicated clearly that its investment in copper will decline dramatically in time and will be replaced by investment in fiber, in digital switching and digital capabilities, and in software. Thus, demand for copper wire is declining rapidly.

Question: Who are the CAPs, and how do they fit into your assessment of the industry?

Comfort: The large CAPs are MFS, which is publicly traded; Teleport Communications Group, which is jointly owned by four cable companies; and (perhaps) Bell Atlantic if the Bell Atlantic–TCI transaction is consummated. Some other regional CAPs and some very small ones also exist. MFS and Teleport have typical strategies based on taking advantage of pricing discontinuities in the access business. That strategy was their entry into the market; they recognized the huge subsidy business customers provide to residential customers. Because they were not fettered by regulation, they could undercut the Bell company or the local telephone company in a region.

This strategy is not a panacea, however, for the CAPs. In time, the Bells will aggressively bring down their access prices and the service will become extremely commoditylike. MFS is valued at multiples in excess of 10–20 times revenue (depending on the day, because the stock is extremely volatile), but to maintain that kind of multiple, it will have to do more than maintain low prices. The real beneficiaries have been the long-distance carriers; they benefitted from the reduction in access charges and are starting to show that boon in their financial results.

CAPs currently have a good and profitable niche in the access business. To be successful in the long run, they need to migrate up the value curve to local dial tone or data services. They must somehow carve out more than just a niche in order to capture the kind of margins that justify their valuations.

The next step for the CAPs, particularly MFS, is to migrate vertically through a product such as data and move to providing dial tone. This strategy will be implemented on a state-by-state basis according to the regulatory environment of each state. In New York, MFS can provide local dial tone and have all aspects of the cocarrier status, and the company has introduced that service competitively. In other states, MFS does not have that flexibility. The regulatory environment will dictate migration to other businesses.

Question: What is the nature of the competition between cellular and SMR services?

Comfort: The whole evolution of the SMR business is interesting. The pace of change and the pace of consolidation in the SMR business put the cellular industry to shame. In less than two years, Nextel Communications increased its population coverage from 60 million to more than 180 million pops (units of population). In contrast, the largest cellular carrier is McCaw with 60 million pops. Thus, Nextel is three times larger, and all this consolidation has happened in less than two years.

The cooperative environment among SMR providers will be important as the SMR service is rolled out and tries to compete with cellular. One of the biggest issues cellular users have, aside from quality of service and capacity in areas like Los Angeles, is that when users roam, they do not know what they will be charged. A New York cellular customer using a cellular phone in Los Angeles does not know the rate of the phone call. The SMRs are trying to provide a seamless roaming capability so that it is transparent in pricing, quality, service, offerings, and every other measure. It is an interesting approach.

Question: What is your interpretation of Nextel's recent acquisition of Motorola's wireless channels in 21 states.

Comfort: The Motorola acquisition can be interpreted in two ways. One is that Motorola is exiting the SMR business by selling out the channels. The second is that it is actually realizing value for its SMR business. I think it is the second. Motorola held the largest asset of SMR properties in the country. It had more than 135 million pops of coverage, but it had two problems. A certain amount of capacity is needed, and the number of channels was insufficient to provide service in a digital SMR platform. The second problem was that Motorola does not like to compete with its customers, which is not a problem for other companies in this industry.

The one way Motorola could have safely realized value is to translate the SMR assets into stock, which is what they did. Motorola traded the assets for equity in customer companies, including Nextel, CenCall Communications, and Dial Page with a two-year lockup (or commitment to holding the stock). It gives them a greater tie to the companies and ensures that the technology is built. It reaffirms the fact that the technology works. In time, this strategy will give the wireless industry a third competitor to contend with—before PCS even takes hold.

Question: What discount rates do you use for international investment to account for the political, repatriation, and government-control risks?

Comfort: We typically use interest rates in the local economy as a starting point and typically adjust for political volatility and stock market liquidity, among other factors.

Question: Since 1991, MCI and Sprint Corporation seemed to have followed AT&T's lead in raising rates, at least for residential and small business customers, and perhaps raising their margins. Is the trend toward higher rates responsible for the run-up in interexchange carrier (IXC) stocks since late 1992? Will this trend encourage policymakers in Washington to remove the long-distance restrictions on former RBOCs?

Comfort: We agree with the observation that long-distance rates have most recently been rising under the price leadership of AT&T. We believe that the IXC stocks have benefitted from a combination of factors, including improved cost structures and growth opportunities from international and data communications, as well as an attractive pricing environment. We would not be surprised if policymakers in Washington use this as a reason to remove Bell RHC restrictions from long distance. Although we expect that over time the long-distance market will open to the RBOCs, we do not believe this will occur without a demonstration of a competitive local telephone market.

Local Telephone Companies

Thomas W. Coyle, CFA
President
Coyle Research, Inc.

> Local telephone exchange companies, which have been very profitable in recent years, provide the bulk of RBOC revenues. They are likely to encounter substantial competition in the future, however, unless they can diversify into wireless and broadband communications.

For the period since the RBOCs began trading on November 21, 1983, through October 15, 1993, all of the telecom companies have substantially outperformed the market (**Figure 1**). Investors who focused on the future and did fundamental analysis of these companies have been well rewarded.

This presentation focuses on some of the fundamental characteristics of the local-exchange companies (LECs) and how investment returns might be affected because of them. The key to future investment success is a function of accurate forecasting of earnings and dividend growth and accurate assessment of managements' capabilities. In many respects, the nonquantitative elements are stronger determinants of future returns than some of the numbers analysts run periodically.

Industry Structure and Profitability

Historically, equity investors in the telecommunications industry had few investment options: With a few minor exceptions, the Bell System, which accounted for 85–90 percent of the industry's revenues, issued common equity through the parent holding company, AT&T. The larger non-AT&T companies (independents), such as GTE Corporation, United Telecommunications, Contel Cellular, and Centel Corporation, mirrored AT&T's practice. Importantly, debt was issued primarily by the operating subsidiaries, such as New York Telephone and GTE California, which numbered more than 50.

When the breakup of the Bell System was being formulated, the parent holding company/operating subsidiary structure was continued. The operating subsidiaries became known as LECs because their business was confined to local areas. The parents became known as regional Bell operating companies or, more accurately, regional holding companies. The RHCs could engage in just about any business except long-distance (interLATA) services, cable TV, and manufacturing. And they have. All now provide wireless services, publishing, and equipment sale and leasing, and they have international operations that include long-distance services. The RHCs have ventured into a number of businesses with limited success or notable failure: financial services, real estate, and computer software and sales head the list of businesses they have entered and exited since their creation.

GTE, Sprint Corporation, ALLTEL Corporation, Cincinnati Bell, and Southern New England Telephone are structured like the RHCs: holding companies with operating subsidiaries that conduct a variety of businesses, the most important of which is providing local telecommunications services through LECs. Rochester Telephone Corporation is an exception to this structure. Rochester Tel's diversified activities are conducted through subsidiaries of the local-exchange company.

The four major determinants of the parent holding companies' profitability are regulators, competition, diversification into new businesses, and new service growth opportunities. Two important areas of analysis for the parent holding companies are the partnerships they will form and the investment opportunities they will have, given the amount of capital they continue to invest in their networks. Analysts and corporate managers must consider how much money needs to be invested in either the local-exchange business or the new business opportunities. Over-investing in the LEC business could result in massive write-offs as competition and technological innovation shorten the economically useful lives of plant. On the other hand, since the

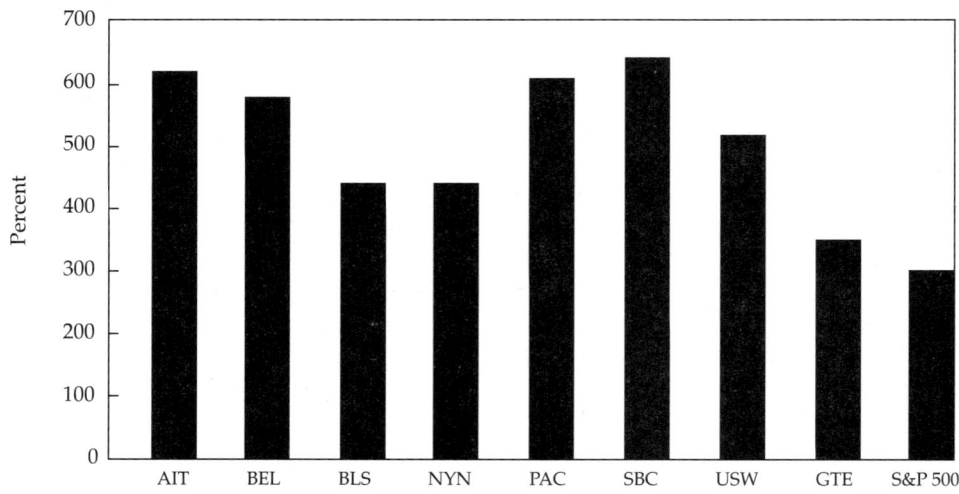

Figure 1. Cumulative Total Return, Telecom Companies and S&P 500, November 21, 1983–October 15, 1993

Source: Ameritech Corporation.

Key:
AIT = Ameritech
BEL = Bell Atlantic
BLS = BellSouth Corporation
NYN = NYNEX
PAC = Pacific Telesis Group
SBC = Southwestern Bell Corporation
USW = U S West
GTE = GTE Corporation

breakup of AT&T, we have seen that diversification can also be fraught with peril.

The LECs are currently very important to their holding companies. Even ten years after divestiture, these companies provide the lion's share of their parents' revenues and operating income, as seen in **Figure 2**. They also account for a substantial portion of the holding companies' identifiable assets, which are revealing because operating profit margins and earnings before interest, taxes, and depreciation (EBITDA) lose their meaning unless they are related to some kind of asset base upon which to calculate return on investment.

The relationship between operating income and assets illustrates that regulated local-exchange operations are much more profitable currently for the holding companies than are nonregulated activities such as cellular and financial services. The identifi-

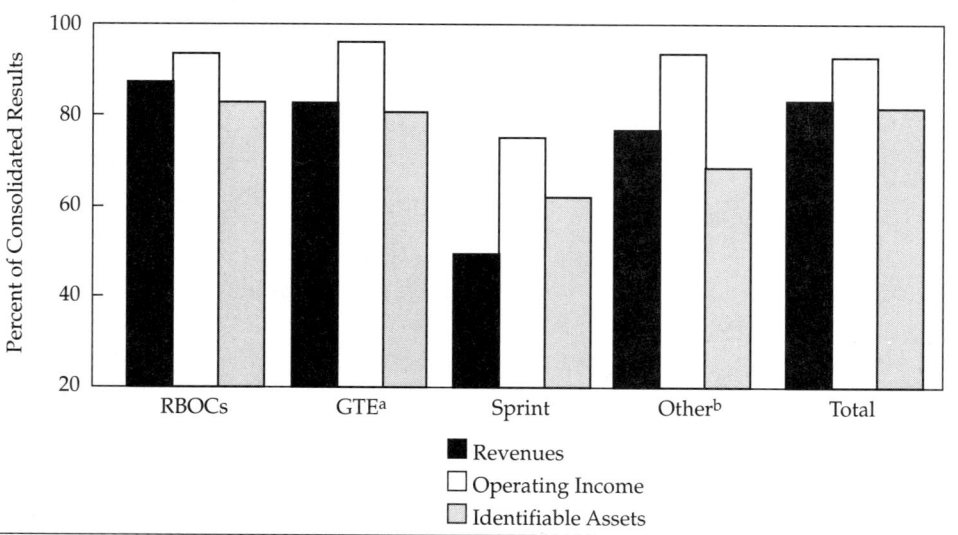

Figure 2. Local Communications Operations, 1992

Source: Thomas W. Coyle, CFA, based on data from company reports.

[a] Average capital rather than assets.
[b] Cincinnati Bell, Rochester Telephone, and Southern New England Telephone.

able assets of LECs probably do not include allocations of parent company overhead assets. If overhead allocations were included, the percentage of identifiable assets attributable to LECs would be somewhat higher, but the relationship between operating income and assets would not change much.

Although the LECs continue as the dominant portion of holding companies' operations, the holding companies' nonregulated activities have grown substantially through diversification since divestiture. The holding companies' non-LEC identifiable assets are $50 billion, revenues are $25 billion, and operating income is about $2 billion.

Regulators influence the financial viability of the holding companies by regulating both interstate and intrastate returns on investment. The estimated returns on the regulated operations shown in **Figure 3** are from company reports filed with the FCC, so they are on a regulatory accounting standard, which is different from the financial accounting standard seen in SEC reports. The returns on intrastate operations and total regulated operations have also been estimated using FCC practices. Federal regulators have one set of rules for determining a company's profitability; state regulators have different rules. The two have many overlaps, but they are not necessarily exactly the same. The interstate returns of 12.2–13.2 percent shown in Figure 3 are as reported by the company and are generally in line with the 12.25 percent allowed under FCC regulation. In most cases, the LECs "share" half of the earnings above 12.25 percent with customers. Returns allowed by state regulators vary from state to state. The intrastate returns of 10.0–10.3 percent are also in line with what state regulatory agencies allow generally, although some consumer advocates have tried to argue that authorized returns should be in the 9–10 percent range.

Full-year returns for 1993 will probably be lower than in the first six months because labor contracts and cost increases come in the second half of the year. Also, the six-month returns on investment for 1993 are lower than the six-month returns on investment for 1992. Both investment and earnings have declined in 1993 relative to 1992, but earnings have been declining faster.

Capital Investment Patterns and Practices

The local telecommunications business has always been capital intensive. As seen in **Figure 4**, three primary categories of plant constitute about 75 percent of gross plant investment. For companies in this universe, cable and wire is the largest portion of the three plant categories, accounting for about 35 percent; for independent telephone companies, which serve less densely populated territories, the cable and wire proportion is almost 50 percent.

Both gross and net investment per line for these three categories of plant vary dramatically by company, as shown in **Figure 5**. NYNEX Corporation and the non-Bells have above-average gross switching equipment investment, possibly reflecting more digital switches. NYNEX went through an accelerated digitalization program, and it has the highest percentage of lines served by digital switches among the RBOCs. The RBOCs and independents have above-average gross and net investment in cable and wire per line. Serving the less densely populated territories requires more capital, but those service territories may not be subject to as much competition as, say, the northeastern part of the United States.

The issue for investors is how much of that investment the companies will fully recover before full-fledged competition emerges. For the purpose of controlling prices for services, in most places, regulators establish the useful lives over which the LEC's investments will be depreciated. **Figure 6** illustrates the number of equivalent years in which the existing investment will be fully recovered. The relationship between the depreciation rate and the number of equivalent years is inverse. For example, a high depreciation rate, such as NYNEX's 6.9 percent for cable and wire, results in below-average equivalent years.

Cable and wire, which is the largest gross value among these investments, has the lowest depreciation rate, which is a major concern for many investors, particularly fixed-income investors. An appropriate policy for a monopoly environment in which no technological breakthroughs are anticipated may be to set low depreciation rates for the largest capital category, but this policy is not appropriate when competition is entering the business and new technologies are emerging.

In September 1993, U S West announced a $5.1 billion write-down of plant investment for financial statement purposes. Figure 6 shows U S West and its regulators have been depreciating cable and wire plant over 21 years. In announcing the asset write-down, U S West plans to depreciate cable and wire over 15–20 years in the future. Similarly, central office switches will be depreciated over 10 years in the future as compared with 18 years previously. Transmission equipment will be depreciated over 10 years also, rather than the previous 13-year life. Fiber optic cable will be depreciated over 20 years rather than 30 years.

Importantly, U S West's decision has no effect on the prices it charges for regulated services. This is because utilities keep two sets of books: one for reporting financial results to investors; the second for

Figure 3. Estimated Returns on Rate Base for Total Regulated Operations, Interstate Operations, and Intrastate Operations

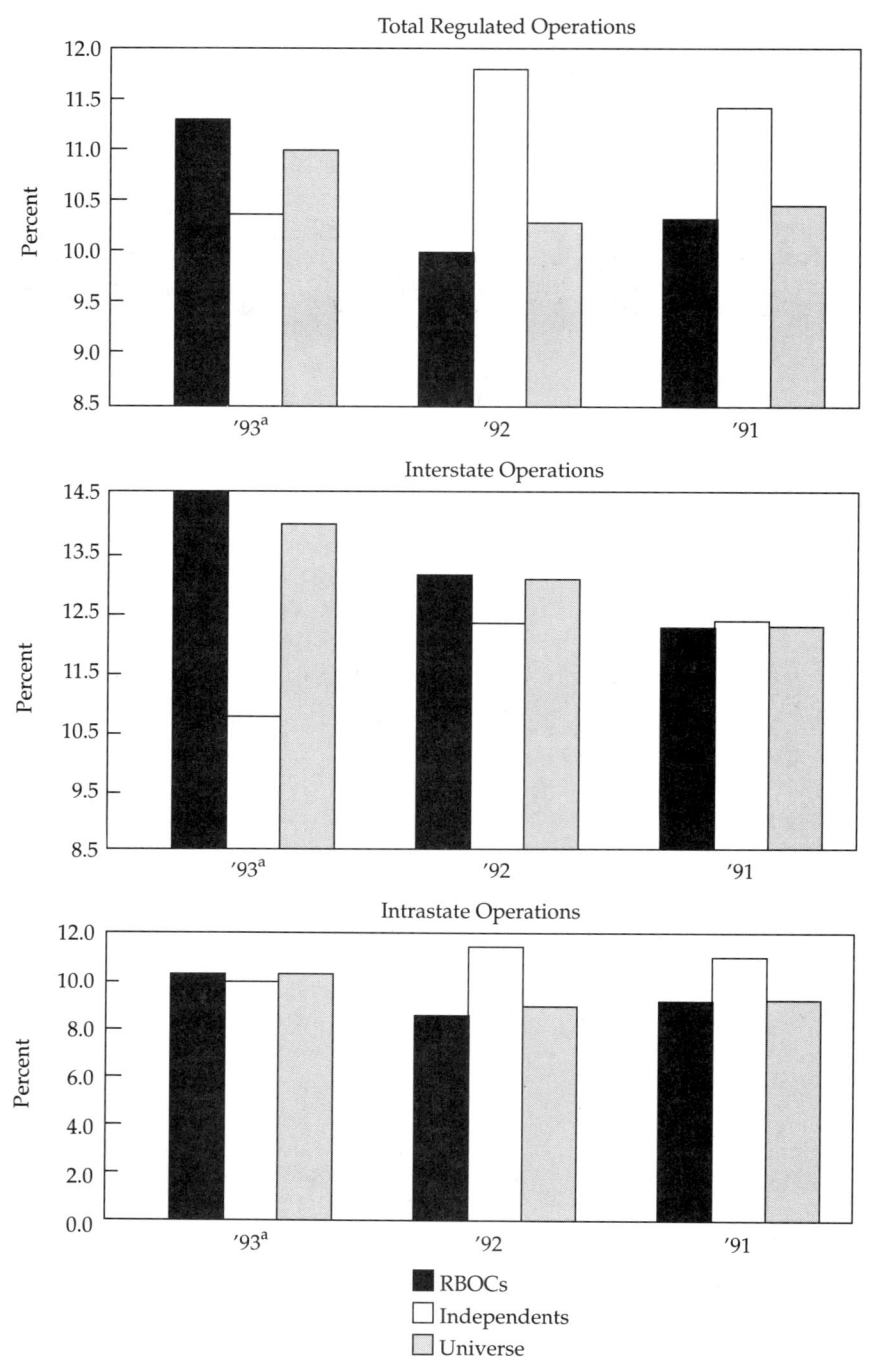

Sources: Thomas W. Coyle, CFA, based on data from company reports to the FCC. Returns on intrastate and total regulated operations estimated by Coyle Research, Inc.

Note: Independents include Centel, GTE, Contel, United, Cincinnati Bell, Rochester Telephone, Southern New England Telephone. Universe includes RBOCs and independents combined.

[a]First six months.

Figure 4. Telephone Plant in Service, December 31, 1992

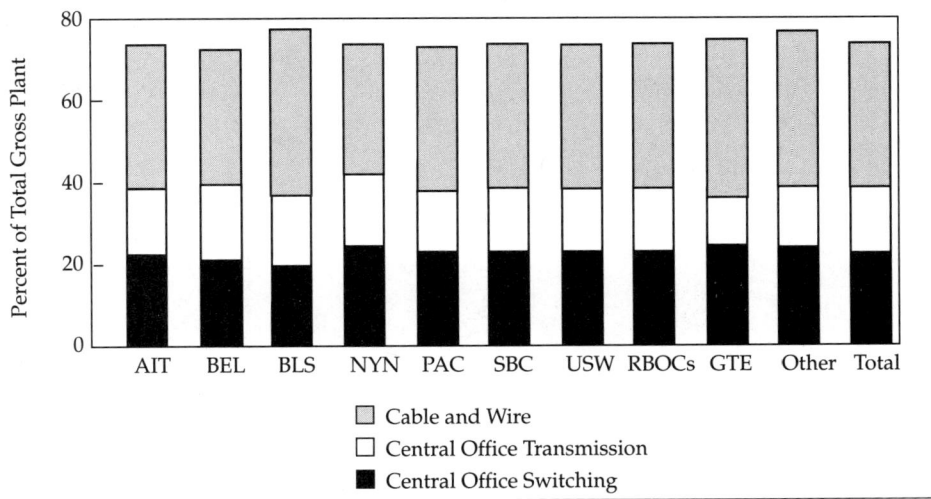

- Cable and Wire
- Central Office Transmission
- Central Office Switching

Source: Thomas W. Coyle, CFA, based on data from company reports to the FCC.

Note: Please see Figure 1 for key to companies.

Figure 5. Investment per Line in Central Office Switching Equipment, Transmission Equipment, and Cable and Wire

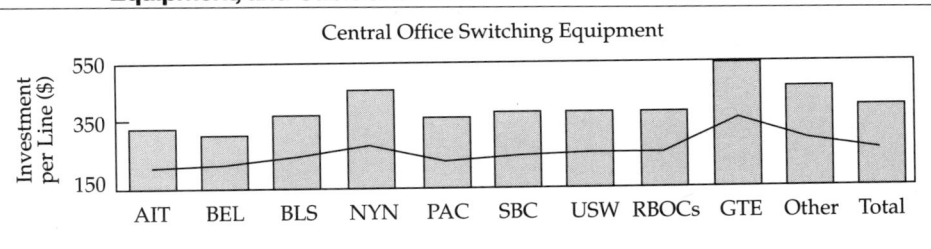

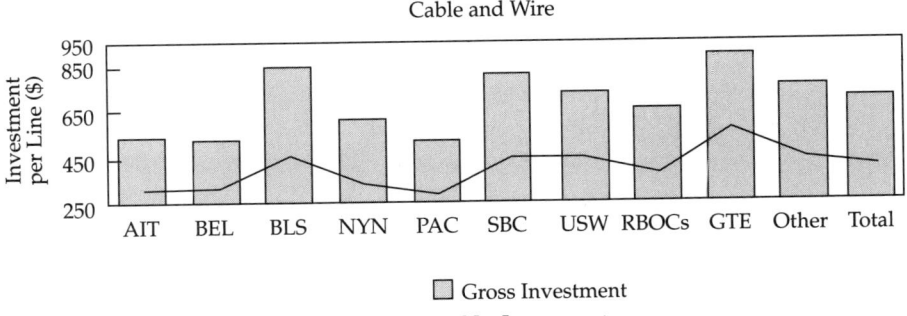

- Gross Investment
- Net Investment

Source: Thomas W. Coyle, CFA, based on data from company reports to the FCC.

Note: Please see Figure 1 for key to companies.

Figure 6. Depreciation Rates and Recovery Periods for Central Office Switches, Central Office Transmission Equipment, and Cable and Wire Plant

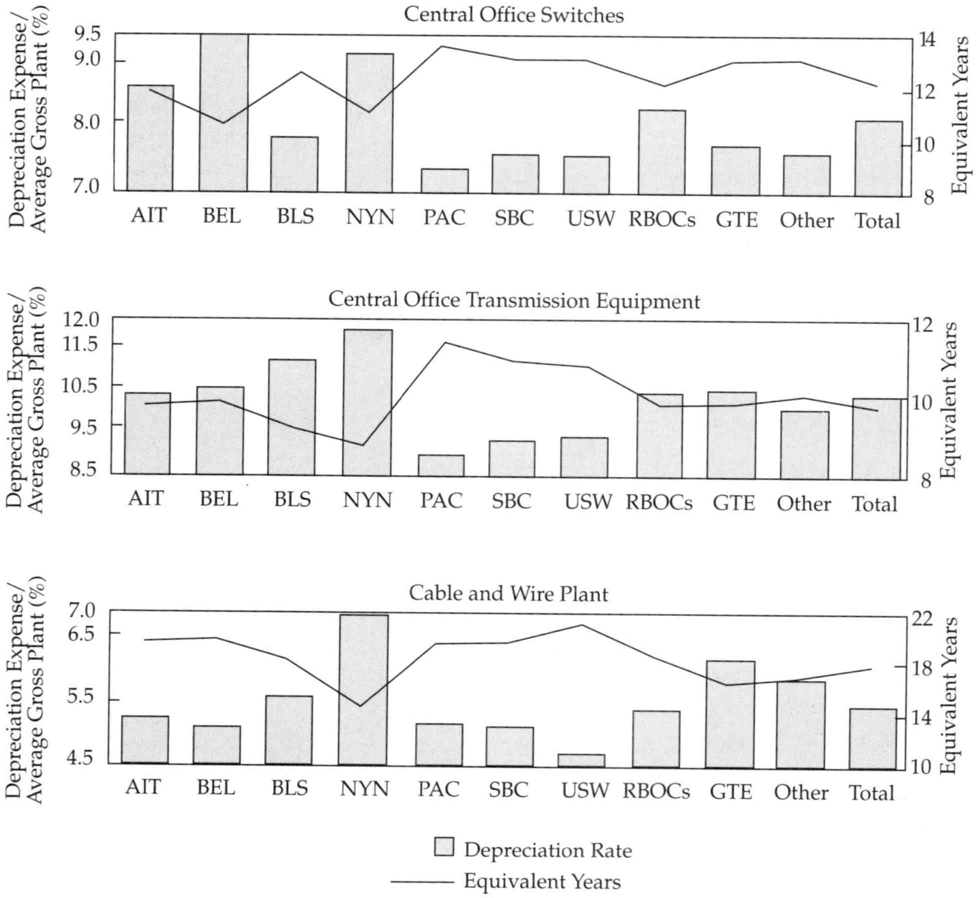

Source: Thomas W. Coyle, CFA, based on data from company reports to the FCC.

Note: Please see Figure 1 for key to companies.

regulators who monitor the company's earnings. For regulatory accounting purposes, the assets have not been written down and U S West continues to have the opportunity to earn a reasonable return on them. For price-setting purposes, U S West will continue to use the average depreciation lives established by regulators.

The announcement had a favorable effect on U S West's stock price. As seen in **Figure 7**, the average closing price for six RBOCs and GTE declined 0.3 percent the day U S West announced the write-down, but U S West's closing price was 4.6 percent higher for the day. Over the entire ten-day period, U S West's cumulative price change was a positive 1.3 percent, while the cumulative decline for the average of the other companies was 3.8 percent.

This market reaction might make managers think that the easiest way to distinguish their stock is to announce a major write-down of assets. This may not be a good long-term strategy, because U S West may begin to report higher earnings and returns on the reduced book value of its common equity. This could draw the scrutiny of consumer advocates and regulators, who for political reasons, may disregard the distinction between financial books and regulatory books and order U S West to reduce rates. This action would reduce regulatory and financial earnings and cash flow.

The biggest benefit of the write-off and the adoption of new shorter useful lives is the effect on U S West's managers, who are responsible for future capital investment decisions. The higher depreciation rates will raise the cost of proposed new investments. This stricter discipline should lead to better investment decisions and fewer and smaller write-offs in the future.

Revenue Sources

Whether U S West and other LECs will be able to recover their current investments through rates for noncompetitive services depends on the competitive outlook. Based on the growth rates shown in **Figure 8**, competition does not appear to be making serious

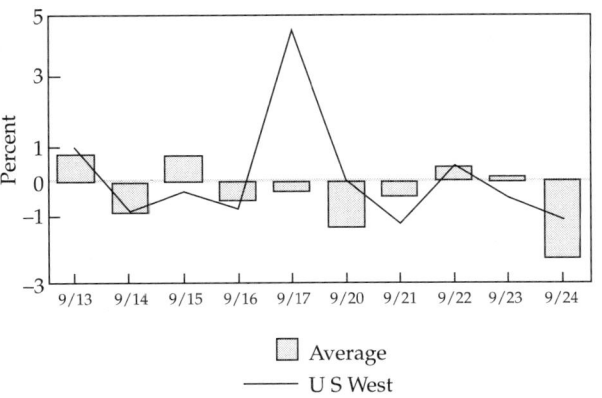

Figure 7. Stock Price Changes, September 13–24, 1993

Source: Thomas W. Coyle, CFA, based on closing prices as reported in *The Wall Street Journal.*
Note: Average includes six RBOCs and GTE.

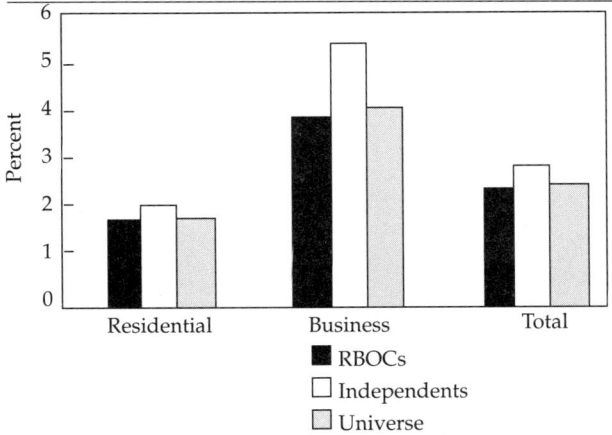

Figure 8. Average Annual Switched Access Line Growth, December 1990–June 1993

Source: Thomas W. Coyle, CFA, based on data from company reports to the FCC.

inroads into their businesses. Although competition exists, and it is growing faster than the LECs, it is not having a large negative effect on a very fundamental measure of growth for LECs. Total switched lines in service mirrored economic activity for the period from December 1990 through June 1993. Residential line growth has been a function of household formation, households taking second lines, and low-income people entering the telephone network. Business line growth has been robust, although with some pockets of weakness in the Northeast and California.

The number of low-income, or lifeline, residential lines increased dramatically during the 1991–93 period, probably faster than the competitors' inroads into the industry.[1] As seen in **Table 1**, for the 35 companies in the ARMIS analyst universe, the average annual rate of growth in residential lifeline access rates was 18.7 percent during this period. For RBOCs, which usually serve urban areas, it was 19.6 percent. At the end of June 1993, 4.2 percent of all residential lines were lifeline lines. This is a significant number, because several large states—including Illinois, Pennsylvania, and New Jersey—have not had residential lifeline services. More than 20 percent of Pacific Bell's residential lines are now lifeline lines. The GTE percentage is about 14 percent, and Contel's in California is between 15 and 20 percent. The rapid growth of these lines in California has resulted in a public utility commission investigation. In a monopoly environment, these lines would not cause a financial hardship. In a fully competitive environment, however, new entrants into the marketplace will fight for the more profitable customers, leaving LECs to deal with this less attractive market segment.

Approximately 30 percent of all lines are business lines. The business access lines shown in **Figure 9** are divided into switched access lines and dedicated access lines (point-to-point lines and lines the long-distance carriers might get from LECs). Call volumes and data-carrying capacities of private or dedicated lines are frequently greater than the equivalent switched access lines. The companies report the types of lines they have in service, but a dedicated line is not necessarily the equivalent of a switched line; it could be the equivalent of multiples of switched lines. The LECs do not report the percentage of revenues from business customers and the percent from residential customers. The general perception, however, is that the revenue split is about 50/50. Thus, because business customers are fewer in number but produce more revenues per customer, the business customer market will be the initial competitive front.

The press reports that access services (the services LECs provide to long-distance carriers) are important to LECs because they produce 25 percent of their revenues and 50 percent of their profits. Data filed with the FCC, however, shows the profitability numbers are substantially different. LECs' interstate access services provide approximately 30 percent of revenues and 31 percent of total LEC profits. The LECs also provide intrastate access services. Those services are probably more profitable than, say, basic residential services, but they are not as high as commonly perceived. Total access service probably represents between 35 and 40 percent of LEC profits.

---

[1] The definition of lifeline varies for each state. Generally, customers must pay a fee, or subscriber line charge, to the local phone company. In most places, this charge is $3.50 a month. If customers meet certain income requirements, the fee is waived. In addition, low-income customers can have telephone service connected at a deep discount and can have their payments spread over a longer period.

Table 1. Lifeline Residential Services

Company	Total Residential		Lifeline Residential		
	Thousands of Lines	1991–June 1993 Average Annual Growth Rate (%)	Thousands of Lines	1991–June 1993 Average Annual Growth Rate (%)	Percent of Total Residential
Ameritech	11,425.1	1.4%	190.8	42.8%	1.7%
Bell Atlantic	11,879.4	1.4	26.8	15.4	0.2
BellSouth	13,474.9	2.8	93.9	217.9	0.7
NYNEX	10,598.5	1.0	765.1	19.8	7.2
Pacific Telesis	9,210.6	2.1	1,864.1	16.2	20.2
Southwestern Bell	8,834.4	2.1	107.4	42.4	1.2
U S West	9,501.4	2.4	229.0	9.7	2.4
RBOCs	74,924.2	1.9	3,277.1	19.6	4.4
Other	14,865.5	2.2	470.8	13.0	3.2
Universe	89,789.7	1.9	3,747.9	18.7	4.2

Source: Thomas W. Coyle, CFA, based on company reports to the FCC.

Keep in mind that LECs are required to use many arcane and improper cost allocation methodologies in reports filed with regulators. Consequently, the profitability of these services, as reported, may be somewhat understated. Finally, even if the LECs lose some of this business, they are unlikely to lose it all.

One way to estimate how much access revenues the LECs are losing is to measure minutes of use (MOUs). Interstate MOUs grew at about the same rate as the general economy from 1991 through June 1993. They did not grow as fast as in the late 1980s, when companies reported per line MOU growth of between 7 and 8 percent. The late 1980s, however, was a period when prices for access services and long-distance services were declining. Thus, the increases in volume were greater than would be seen in a stable price period. **Figure 10** reflects the growth of interstate MOUs during a period of fairly stable long-distance prices.

Analysts can get a sense of how much bypass is occurring by monitoring interstate originating MOUs.[2] The originating MOU is the opposite of the terminating MOU. If the LECs were handling all calls, 50 percent of the MOUs would be originating and 50 percent terminating. As shown in **Figure 11**, originating MOUs fluctuate between 42 and 45 percent with a small downward trend of probably less than 1 percent during a period of 2.5 years. This imbalance between originating and terminating MOUs provides an indication of the degree to which the LECs' networks are being bypassed on the originating end of calls. Thus, even during the period from 1991 through the first six months of 1993, when little competition existed, the LECs were losing originating MOUs.

---

[2] An originating MOU is a minute of use on the originating end of a call.

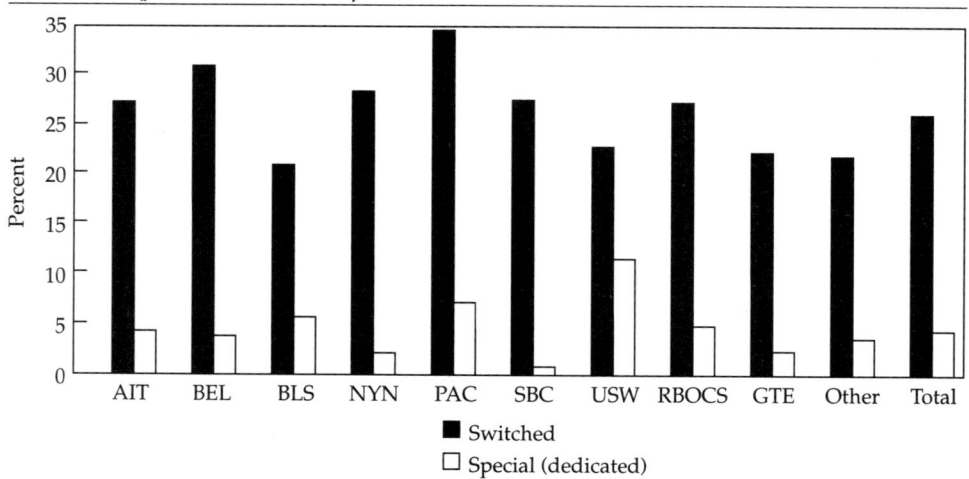

Figure 9. Business Lines (percent of total lines)

Source: Thomas W. Coyle, CFA, based on company reports to the FCC.
Note: Please see Figure 1 for key to companies.

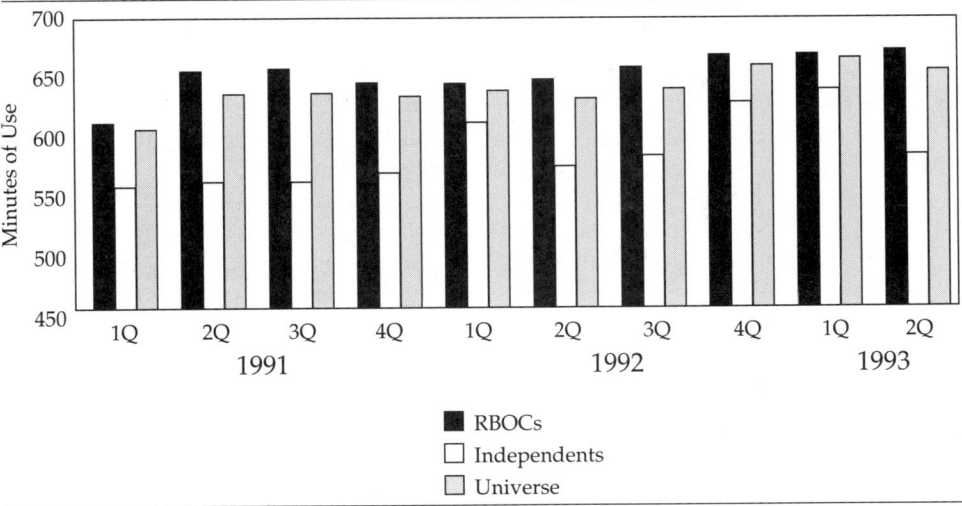

Figure 10. Interstate Minutes of Use per Line

Source: Thomas W. Coyle, CFA, based on company reports to the FCC.

Competition Potential

So far, competition has not had a serious negative impact on the volume measurement of LECs' businesses, but competition from CAPs, cellular, and cable television will affect them in the future. Currently, one cellular line is in service for every three switched business lines. At some point, business people will decide they do not need two telephones when one will do. New cellular customers are marginal customers and marginal users. Many of the new customers subscribe to off-peak pricing plans. With cellular lines growing 25 percent a year, at some point, cellular companies will be able to charge customers two different rates. They will charge customers calling another cellular customer one price, and they will charge customers calling a wireline (LEC) customer the cellular customer rate plus what they have to pay the LECs. This policy would give customers a large incentive to switch from wireline to wireless service. The cellular companies say publicly they are not interested in this strategy, but at some point, it will make sense for them.

The CAPs are another growing competitive sector. They began by providing dedicated fiber optic communications facilities, but they are beginning to deploy switches. One CAP's switch turned on less than two months ago handled 600 calls on its first day of operation—not very much for millions of dollars of investment. On the 45th day, however, it handled 30,000 calls. The CAPs will begin to roll out switching services because the revenue payback from offering switched services is much greater than it is from point-to-point services.

Universal service is another issue of future concern. New competitors—CAPs, wireless carriers, cable TV companies—do not want low-income lifeline residential customers. In most cases, they talk about paying to have somebody else provide low-income services; they are willing to contribute to a high-cost or subsidy fund, but they want LECs to handle lifeline services for them. That same pattern will happen with universal service. At some point, however, a battle over the allocation of costs for those universal services is inevitable. Either the regulators will be under pressure to disallow some of the costs, or the CAPs will find a way to get around them.

The current rage in the industry is cost-based pricing. The problem LECs face as they move to cost-based pricing is that they are shifting more of their cost burdens onto residential customers. At some point, the competitors are going to go after that residential customer base.

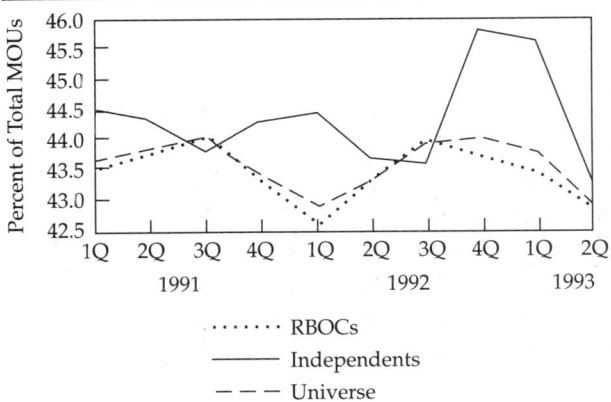

Figure 11. Interstate Originating Minutes of Use

Source: Thomas W. Coyle, CFA, based on company reports to the FCC.

Conclusion

In summary, investors who evaluated the fundamentals of the telecommunications industry have been well rewarded in the decade since the breakup of AT&T. Attention to company fundamentals will be more important in the years ahead. Local-exchange operations continue to be the dominant portion of most companies and will deserve the most attention in the next five to ten years. These operations are currently heavily regulated and face a competitive future. Competition has not had a significant effect on the LECs thus far but will be a more important factor in the future. As competition increases, the LEC subsidiaries could face major asset write-downs, as well as deteriorating profit margins and cash flow. A major risk is that regulators will not give the LECs sufficient freedom to respond to competition. Debt investors have the greatest risk because the LECs, the primary issuers in this industry, face the greatest business restrictions. The holding companies have less risk because they have fewer restrictions. The risk of the holding companies lies with their diversification decisions. Bad diversification decisions, coupled with deteriorating LEC cash flows, on which the holding companies are heavily dependent, could restrict the holding companies' growth prospects.

Question and Answer Session

Thomas W. Coyle, CFA

Question: Relying on your background as a credit analyst, do you have any insights into capital structure and costs? What is the optimal capital structure in light of the expected future of the industry?

Coyle: U S West's capital structure will be considerably different from what it was before the writedown. The RBOCs would be making a mistake if they deliberately overleverage their companies at this point, because the odds of seeing cable and wire write-downs in the future are significant, and those write-downs will result in a more highly leveraged capital structure than in the past. As competition enters this industry, profit margins will deteriorate somewhat across the board, not just for access services. This industry has generally had Aa credit ratings, but over time, these ratings will probably decline to A, A+, or even some Bbb ratings. A large part of this decline will be a function of the regulatory environment, not necessarily of poor management strategy.

Question: Are you aware of any tentative positions state regulators have taken in light of the Bell Atlantic Corporation–TCI merger plans?

Coyle: No, but frequently, at conferences, I hear about the RBOCs' use of funds. Regulators perceive that the diversification activities these companies have embarked on have come out of the pockets of their customers. They are correct in one respect; when customers pay their telephone bills, part of that payment applies to the recovery of capital (depreciation of plant). Management's obligation is to determine the smartest place to redeploy that capital. Management seems to think it makes more sense to deploy the capital outside the voice-grade, wired telephone business and enter broadband or wireless activities.

Question: How do credit analysts perceive the planned mergers between cable and telephone companies?

Coyle: Credit raters focus on cash, particularly who has the cash and who is getting the cash. When cable TV companies leveraged up in the last part of the 1980s, they said they were going to cover their interest expense, that is, their EBITDA coverage of interest expense was going to be 1–1.5 times. They planned to use the capital recovery, or depreciation, money to pay fixed-income investors' interest payments. That strategy was reflected in their poor credit ratings. The rating agencies believed that this strategy was not very viable long term because they should pour the depreciation money back into the network. The reason for some of the mergers that are occurring is that the cable TV companies need the cash flows from the telephone companies. If they were not so highly leveraged, they might not need that cash flow so badly.

Question: Do you see the need for credit rating agencies to adjust rating standards to address increased competition, as they have for electric utilities?

Coyle: I believe that competition will be a more important factor in the future. I also believe the telecommunications industry is closer to meaningful competition than the electric utility industry. New York Telephone's current single-A ratings reflect a combination of the attractiveness of New York Tel's market to competitors, as well as the company's below-average financial performance. I believe the credit rating agencies are very aware of the exposure the LECs have to competition, but to date, the effects of competition have not been significant.

Competition in the Alternative Access Market

Mark Lowenstein
Associate Director
The Yankee Group

> Traditionally offering local-exchange bypass at a discount, CAPs are now moving toward providing full telecommunications services. With backing from cable companies and an improving regulatory situation, CAPs could grow significantly.

Alternative access carriers, also known as competitive access providers, are attracting attention among investors, not because of the profits they earn, but because of the potential they offer. At the end of 1992, about 35 major cities had CAP networks. By the end of 1993, about 40 of the top-50 metropolitan statistical areas had CAP networks. Increasingly, many secondary cities also have CAP networks.

The CAP Market

CAP networks are usually fiber networks in major downtown or business/industrial park areas. The primary business of CAPs is to provide direct connections from large business customer premises to an IXC's point of presence (POP). They bypass traditional local access networks, connecting the customer's premise directly to the long-distance network.

Some CAPs such as Locate in New York provide digital microwave-based bypass. Although 75 percent of the business of CAPs is connecting customer premises to a carrier POP, they also service some IXC's POPs in the same city. For instance, if AT&T has POPs at either end of the city, sometimes it uses the CAP network to take traffic from one POP to the other. The CAPs provide some customer premise-to-premise business, as well.

When the CAPs initially entered the business, they had two primary selling points. They offered 20–30 percent lower prices than the traditional local-exchange carriers for dedicated access lines—the traditional high-speed T-1 lines. That price gap narrowed, however, when the RBOCs succeeded in lowering their private-line and dedicated-line tariffs. Another reason CAPs were an initial success is because they had redundant fiber networks. The LECs did not have those types of networks in place, and many users wanted alternative networks to use in addition to their LEC networks. Thus, many customers in the financial community used some CAP network facilities and some LEC facilities. In fact, very few CAP customers use a CAP network exclusively, and the same is true for the IXCs.

The geographic layout of CAPs changes daily, but as seen in **Exhibit 1**, as of mid-1993, CAPs had operational networks throughout the country. Twenty cities have projected networks that are financed and under construction. Some of the projected networks have actually received public utility commission permission and are actively building networks. Many cities such as Chicago, Boston, San Francisco, and Los Angeles have multiple CAP networks, an increasing trend.

Industry Growth

Obtaining exact financial information on CAPs is very difficult. Many of them are small, privately held, or owned by cable companies that do not break out the revenues of their small subsidiaries.

The CAP market is still relatively small, and it has experienced only moderate growth. At the end of 1992, it amounted to just under $200 million in revenues, and it grew to about $210 million in 1993. As seen in **Figure 1**, growth projections for 1994 and beyond are more aggressive. One reason is that three dominant CAPs are actively building networks and adding to their own market portfolios. They are starting to develop sustainable revenues and a track record.

A second reason for higher expected growth is that CAPs can now compete in the switched services market. Until recently, CAPs could only compete directly with the RBOCs in private-line, or dedicated, services. They could not install switches or offer

Exhibit 1. U.S. Competitive Access Provider Network Locations

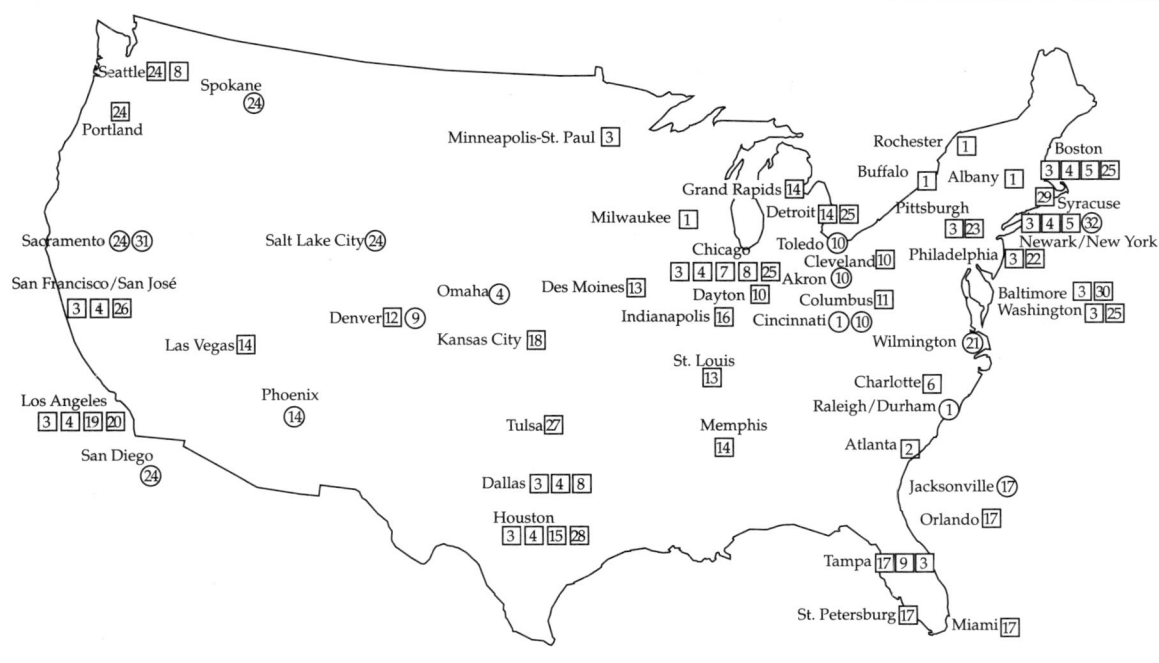

Alternative Access Carriers

1 = FiberNet (Time Warner)	9 = Jones Lightwave	17 = Intermedia	25 = LOCATE
2 = Metrex (MFS)	10 = OhioLinx (Teleport Denver)	18 = K.C. FiberNet	26 = Bay Area Teleport
3 = MFS	11 = MetroComm (Time Warner)	19 = ACLA	27 = Tulsa Metrolink
4 = Teleport	12 = Teleport Denver	20 = MTel	28 = Houston Power and Light (MFS)
5 = New England Digital (MFS)	13 = MWR Fibercom	21 = Delaware Lightware	29 = New Channels Hyperion Telecommunications
6 = Privacom	14 = City Signal	22 = Eastern Tele-Logic	30 = Baltimore Utilities (MFS)
7 = Diginet (Teleport)	15 = Phonoscope (Teleport/Digital Direct)	23 = Penn Access (Digital Direct)	31 = Phoenix Fiberlink
8 = Digital Direct (Teleport)	16 = Indiana Digital Access	24 = Electric Lightwave	32 = MH Lightnet

☐ Operational
○ Projected

Source: Yankee Group, 1993.

switched services. Now, however, the FCC has ruled that the LECs must provide CAPs with switched access interconnection. This ruling allows CAPs to co-locate in LEC switches and offer switched services. Issues remain about state utility commission involvement and whether, despite the FCC's ruling, the full force of the change will occur. Investors will see the revenue impact of the switched services ruling as the CAPs become full-service providers. As a result, this market will become significant during the next couple of years. The states may become more aggressive in allowing competition.

Some of the first CAP trailblazers were Metropolitan Fiber Systems (MFS) and Teleport Communications Group in 1985. Growth has been rapid in the past few years, particularly in 1993, with a lot of new construction. CAP networks will be available in almost 60 cities by the end of 1993.

The competitive situation in certain cities has led to a market shakeout during the past 12–18 months. As seen in **Figure 2**, the number of cities served by CAPs has increased even though the number of CAPs has decreased. The industry is beginning to consolidate, and even though construction continues to blossom and revenues continue to grow, the number of providers is shrinking. Some of the larger CAPs, such as MFS and Teleport, rather than building a competing network, are buying CAPs in smaller cities to add to their portfolios. Large cities, such as New York, Chicago, and Boston, can support more than one alternative fiber network, but they cannot support three or more plus the LEC that has been aggressively deploying its own fiber as a competitive response. Some of the most financially successful CAPs are those that have entered the secondary cities as sole providers. Those in smaller cities, such as Kansas City and Omaha, have been fairly successful because another CAP is unlikely to

Figure 1. Revenue Growth Model of the CAP Market

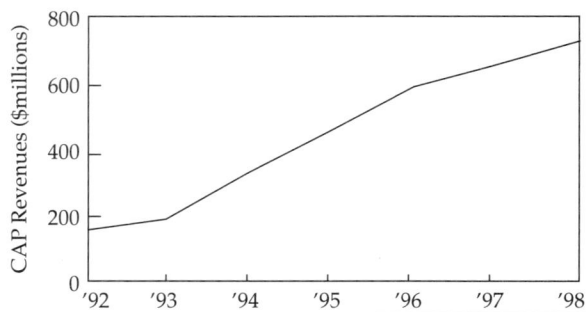

Source: Yankee Group.

Table 1. Network Size of CAPs

Competitive Carrier	Number of Networks	Route Miles	Buildings Served
Teleport	10	650	850
MFS	13	500	850
TWE AXS	7	350	325
Intermedia	4	165	105
15 other carriers combined	26	405	105
Total	60	2,070	2,235

Source: Yankee Group.

enter into competition there. Also, those cities are not high on the RBOCs' priority list as places to build fiber networks.

CAP Networks

Route miles and fiber miles are two distinct measures of company networks. Many companies disclose fiber miles because that figure looks incredibly impressive, but the important measure of network size is route miles. As seen in **Table 1,** Teleport has ten networks and 650 route miles. Most competitive access networks average 5 or 6 route miles in any given city where they operate. Some anomalies exist, such as Eastern Tele-Logic, which provides the network to the Delaware Valley. Teleport's network in this area is very extensive because it includes Manhattan and New Jersey operations, but generally, the individual networks are fairly small and only cover the critical mass of buildings. Fiber transmission must be built directly out to each customer, which is an expensive proposition. MFS, which went public in May, has 13 networks serving about the same number of buildings as Teleport. Intermedia Communications, the other public company in the business, has four networks covering the main cities in Florida—Orlando, Tampa, Miami, and Ft. Lauderdale/West Palm Beach. The other 15 carriers combined have a total of 26 networks.

The total number of route miles becomes significant considering that more than 2,000 route miles of fiber have been built. In aggregate, that amount approaches almost a coast-to-coast long-distance fiber network. Obviously, the network configuration is not coast to coast, but that much mileage represents a lot of money invested in this business. We estimate some $700 million to $750 million has been invested in CAP networks since the mid-1980s. The RBOCs have also invested heavily, bringing the total invested in metropolitan area CAP-type networks to $1 billion.

Profitability Factors

Among the success factors in this market is being first to market. CAPs have been most successful where they are the only ones in the market—some of the secondary cities. That is one reason Intermedia has been successful in Florida. It has not had much competition from other CAPs and only a delayed response from the competing RBOC. The CAPs that have been most successful financially are Teleport and MFS, which were institutionally backed. Teleport was initially backed by Merrill Lynch, which later sold its interest to the cable companies. Before it went public, MFS was owned by Peter Kiewit Sons, a large construction firm with various holdings. These two companies did not have to rely on venture money. As a result, they enjoyed the luxury of building networks without needing to show quick profits. The financing enabled them to make the necessary investments to become important forces in those markets.

To date, CAPs have not been very profitable. Although Teleport has turned a profit for most of the

Figure 2. Historical Timeline of the CAP Industry

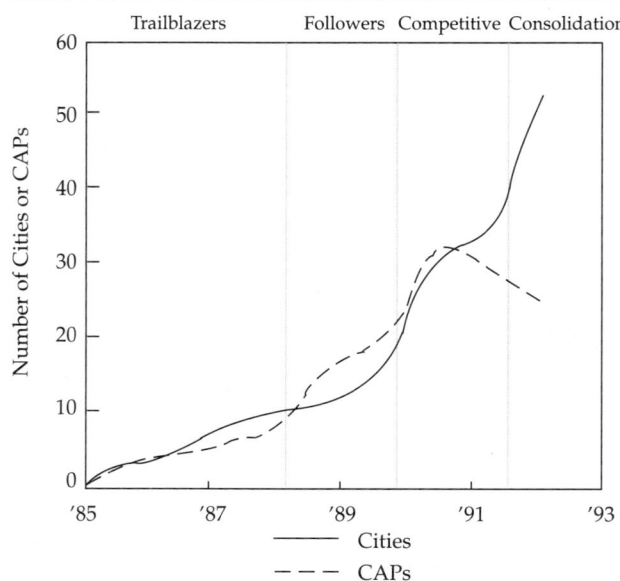

Source: Yankee Group.

years of its operations, MFS has not. The industry's potential has had more impact than its revenues or profits. The price of MFS and Intermedia on the markets indicates that many investors are excited about the potential of these companies despite the lack of profits.

Certain factors influence CAP profitability and will continue to do so in the near term, as shown in **Table 2**. One negative impact is that the RBOCs continue to erode their prices for DS1 and DS3 access (very high-speed private lines) in competitive zones. The utility commissions have given the RBOCs certain pricing flexibility in exchange for sharing their switches with CAPs. As a result, the RBOCs have the luxury of being able to reduce prices in select markets; if CAPs reduce DS1 prices in New York City, they have to do it throughout the state of New York.

Another negative impact on CAP profitability is that some of the RBOCs are charging fairly high interconnection rates to CAPS that want to enter the switch business. This pricing policy has caused battles between some of the CAPs and RBOCs. MFS, for example, recently filed a letter opposing the proposed Bell Atlantic Corporation–Tele-Communications (TCI) merger. One reason is that MFS sees Bell Atlantic's interconnection charges as more unfair than those of other RBOCs.

In other areas, CAPs and RBOCs are cooperating based on the realization that deregulation for the RBOCs is inevitable. NYNEX Corporation, for example, has been fairly even-handed in its dealings with Teleport as it negotiates to use NYNEX switches. The CAPs' $200 million in access revenues is a small piece of the RBOCs' business, and the RBOCs have many other things to be concerned about from a competitive standpoint.

Positive issues affecting CAPs include the regulatory ability to get into the switch market; favorable rulings on interconnection; and increasing involvement with cable companies, which provide them with capital for expansion and help build the infrastructure necessary to serve more businesses and expand the range of their businesses.

A potential problem is that the IXCs are eager to build their own CAP networks. A few years ago, MCI Communications Corporation purchased a subsidiary of Western Union called Advanced Transmission Systems (ATS), which has fiber networks, some of which are lit, or active, and tremendous right of way dating back to 1880 in cities all over the country. To date, MCI has not done much with that asset, but it has the potential to build CAP networks of its own in some cities so it could offer direct connections to its network from customer premises. This is now the key strategy behind MCI's metro plan.

RBOC Competition

To date, RBOC revenues have not been significantly affected by competition from CAPs, despite the RBOC filings with the FCC about revenues lost to bypass. The $200 million industry is not even 1 percent of the entire RBOC access revenues. In select markets such as Boston or New York, however, and in particular markets where CAPs have been especially successful in downtown business access, some revenue erosion has occurred.

Much of the hubbub about CAP competition has been because of the potential erosion of RBOC revenues as the CAPs are allowed into more lines of business that will compete with the RBOCs and as these networks become more pervasive in size within a city and in the number of cities where they operate. As seen in **Table 3**, some RBOCs face competition from more networks, and those will be affected more than others. Ameritech Corporation, NYNEX, and Bell Atlantic lead in the number of competitive access networks projected or operating

Table 2. CAP Profitability Factors

Issue/Event	Timing	Potential Impact
Further rate decreases in competitive zones for DS1 and DS3	2nd/3rd quarter 1993	Negative (moderate)
High cross-connection charges for interconnect. Special and switched	2nd/3rd quarter 1993	Negative (significant)
Switched access transport declared competitive—rates greater or equal to special access DS1 and DS3 rates.	3rd/4th quarter 1993	Positive (significant)
Regulatory ability and marketing success in given states re: local switched services	1994–95	Positive (moderate – significant)
Ability to provide cable TV and telephone services as a single provider	1993–94	Positive (moderate – significant)
Political or other factors which slow down the process for pro-competitive action/services	1993–94	Negative (significant)
Major IXCs building their own interconnections as opposed to leasing from CAPs	1993–94	Negative (significant)

Source: Yankee Group.

Table 3. CAP Networks in Place/Projected by RBOC Region

Company	Total	Operational	Planned
Ameritech	15	12	4
Bell Atlantic	11	9	2
BellSouth	9	5	4
NYNEX	11	8	3
Pacific Bell	10	7	3
Southwestern Bell	9	9	—
U S West	7	6	1
Total	73	56	17

Source: Yankee Group.

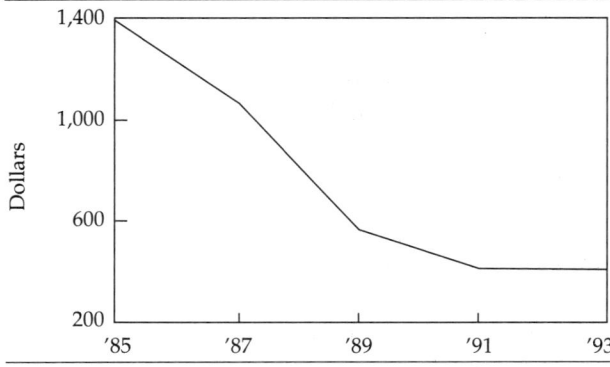

Figure 3. Average Price per Month for a T-1 Circuit

Source: Yankee Group.

in their regions.

To respond to the perceived competitive threat, the RBOCs are offering leading-edge services that the CAPs cannot yet provide, especially in those states that prohibit CAPs from offering switched access interconnection. Also, the RBOCs have started spending heavily on fiber networks. As **Table 4** shows, Ameritech, Bell Atlantic, and NYNEX have been among the more aggressive RBOCs, probably because they face the greatest threat from CAPs. They have made substantial investments in deploying fiber rings of their own. The cost estimate to put fiber into a region is based on fiber that is already in place and the geography of the region. Table 4 also lists other business the RBOCs are entering.

RBOCs have also been able to compete by lowering their prices. As seen in **Figure 3**, the average monthly price for a T-1 circuit dropped significantly between 1985 and 1991 and has since remained relatively stable. Since 1991, the arena of competition between the CAPs and the RBOCs has shifted from price toward customer service. The RBOCs are focusing on issues such as the time required to deliver a circuit to the customer, bit error rates, network availability, and the number of services offered. Price is no longer the big selling point because a 5 or 10 percent cost savings is not a big deal for the customers that use these alternative networks. They are willing to pay a significant amount of money for redundancy. CAPs are the preferred choice for users' disaster recovery plans. In some cases in which the CAPs charge more for services than the RBOCs, customers still want the diversity or redundancy the CAP offers.

With the grounds for competition shifted away from price, as CAPs consider taking on other lines of business and as the types of services converge, the whole industry is at an important turning point. Diversity and redundancy are still major issues among large and medium-sized business customers, especially as communications become increasingly critical. The original competitive advantages for the CAPs—first to market and lower price—have started to erode, and they have a whole new outlook focusing on where they can go next and what they can do to stay in business, be competitive, and develop revenues.

An important development is that the regulatory environment is changing. The FCC has been moving in the direction of allowing switched access, which will get the CAPs into the switch business. That ruling, however, is still more of a signal than anything else. State commissions still have some control over competition within their jurisdictions. Just because the FCC has made this ruling does not mean the CAPs can start putting in switches and offering switched access services nationwide. They must get permission from the state commissions, and they

Table 4. RBOC Competitive Response Analysis

RBOC	1992 Capital Expenses (billions)	Miles of Copper (millions)	Miles of Fiber (millions)	Estimated Cost to Fiber Region (billions)	Comments
Ameritech	$2.3	191	0.59	$14.75	Testing video on demand and educational systems
Bell Atlantic	2.2	197	1.20	30.00	ADSL, 100% F/O in NJ, Video on demand
BellSouth	2.8	122	0.94	23.50	Video via F/O, Cable TV, and phone tests
NYNEX	2.5	167	0.97	24.20	Liberty Cable test, UK Cable
PacTel	1.7	156	0.31	7.70	100% F/O proposal by 2015
Southwestern Bell	1.6	170	0.58	14.50	Cable television in Virginia
U S West	2.4	167	0.80	20.00	Time Warner deal, TCI/AT&T Video on demand trial

Source: Yankee Group.

must negotiate interconnection agreements with the RBOCs, which will be a steady but slow process.

The most progressive states, New York, Illinois, and Michigan, allow full-fledged competition in both private line and switched services. Massachusetts and Oregon are on a fairly fast track to resolve the issue. Texas and California have the issue on their dockets and consider it fairly important. Looking at the summation of all the CAP networks, these states tend to be the ones where several networks exist and where competition in the whole communications infrastructure is important from the perspective of retaining businesses.

Cable Investment in CAPs

Through either direct ownership or backing, cable companies influence an estimated 50–60 percent of CAP revenues. The extent to which cable companies are becoming involved in the competitive access business is shown in **Table 5**. CAPs were the cable companies' first entrance into the telecommunications business. This involvement was the first real signal that the cable companies are determined to get into the telecommunications business, that they see a role for their networks in the communications infrastructure beyond just serving customers with cable news and ESPN. This horizontal integration will affect the competitive access business and change the traditional models of competition and regulation between cable and telephone companies. The impact of cable investment has been that the CAPs have the luxury of a longer term view of network construction, profitability, and the services they can offer, because they have more solid financial backing. The cable investment tends to encourage metro area networks (larger networks within cities). The cable infrastructure also serves networks interconnecting cities and offers new services such as regional interconnection. CAPs and cable companies have even cooperated in some of the PCS trials. The CAPs are starting to evolve beyond just providing a pipeline between a customer premise and a carrier POP in all of these new services. That expansion will probably be the most exciting development during the next few years.

Future Developments in Telecom

The competitive access industry will probably consolidate into three national CAPs with networks in multiple cities across the United States. These firms will be Teleport, MFS, and the U S West–Time Warner entity. The latter has only four networks right now, but Time Warner owns or backs four CAP networks in different parts of the country. As a result of the capital infusion, the company has announced plans to build more networks. U.S. West has no qualms about building competitive networks using Time Warner facilities to compete with RBOCs in

Table 5. Horizontally Integrated Cable Companies, Top-25 Multisystem Operators

Cable Company	CAP Ownership or Backing	PCS Trials/ License Application	Owns Cellular Properties
Adelphia Cable	√	√	
ATC	√		
Cablevision Industries		√	
Cablevision Systems		√	
Cencom Cable		√	
Century Communications	√		√
Comcast	√	√	√
Continental Cablevision	√	√	
Cox Cable	√	√	√
Falcon Cable			
Jones Intercable	√	√	
KBLCOM	√		
Lenfest Group			
MultiVision Cable Television			
Newhouse Broadcasting			
Paragon Communications	√		
Sammons Communications			
Scripps Howard			
Storer Cable			
TCI	√	√	√
Tele-Media			
TeleCable		√	
Time Warner	√	√	√
Times Mirror Cable Television			
Viacom Cable		√	

Source: Yankee Group, 1993.

cities outside of its home service area. Management clearly stated that strategy when the deal was originally announced.

CAP networks will continue to focus specifically on business customers with switched and dedicated services. CAPs are not eager to get into the residential market. It is not a very profitable business and must be subsidized by the LECs. The CAPs might get into residential services other than telephone, but they will use cable companies' assets to do it.

Fiber-based CAPs are in more than 50 markets now, and the RBOCs are facing competition for their larger business customers. The long-distance carriers have endorsed the CAPs and are using their networks in many cities. AT&T is an aggressive user of alternative networks and actively promotes alternative networks in some cities. The regulators are allowing these competitors to expand into more lines of business. Cable companies are entering the business and some are involved in wireless as well. Companies such as Locate that offer digital microwave-based bypass are also running PCS trials. That raises the possibility of complete bypass of the local network using wireless local loops, one of the key promises of PCS. One aspect of PCS that could differentiate it from cellular is the prospect of complete and wireless networks.

Wireless

CAP networks have been used in several cities to carry traffic from cellular-based stations to carrier POPs. For example, Comcast Corporation uses Eastern Tele-Logic in Philadelphia to provide direct connection between customer PBXs and carrier POPs. These CAP networks have become important to the wireless business, as well. Cable companies have been one of the most aggressive pursuers of PCS. They see PCS as an opportunity to get into wireless, and we think they have been involved in many of the trials and market tests. They will be very aggressive in pursuing PCS licenses now that the FCC has ruled on the PCS licensing process. MCI is determined to pursue a national PCS license, but it will need deep pockets and help to do it. MCI will probably bid for spectrum along with cable company partners in its effort to get into the wireless business.

Wireless is the fastest growing area in the communications industry: Existing cellular is growing 20–25 percent a year. In comparison, local-exchange business is growing 4–5 percent a year in access lines and long distance will grow 7–8 percent a year during the next few years. Enhanced specialized mobile radio (ESMR), a new type of cellular service, is growing faster than regular cellular. One ESMR company, Nextel Communications, frequently appears in the news as the subject of a new deal to buy out other companies and aggregate markets.

The value of spectrum is growing in importance. The companies Nextel is buying are very small and uninfluential, but the deals have been highly valued because spectrum is scarce and valuable. The Clinton Administration decided that PCS spectrum is worth about $10 billion, so acquiring it will take some deep pockets.

Realizing the potential threat from AT&T–McCaw, the RBOCs have formed an alliance, called MobiLink, made up of their B-side (wireline) cellular operations. The purpose is to provide seamless roaming and other services on the cellular network, better customer service, service guarantees, and so forth, mainly to compete with what they perceive to be a national cellular–IXC bundled service AT&T–McCaw could provide. All of these elements are bringing dynamic change to the communications industry, and they have the potential to alter completely the traditional model of the LEC business.

With Pacific Telesis Group spinning off its wireless PacTel division and with alternative networks pressing to expand into more services, the regulatory structure of the LEC business will change during this decade, and probably sooner rather than later. The RBOCs have ample cause now to argue to the FCC for greater freedom from some of the modified final judgment restrictions. The FCC has been lenient in ruling favorably on applications for interLATA (long-distance) service waivers for RBOCs' cellular businesses so they can offer roaming in contiguous service markets that happen to be interlata. That is a good signal of loosening regulation of the traditional regulated business.

Cable

The RBOCs now have a cable strategy. They are entering the pay-per-view and video-on-demand business, and cable and telecom alliances are taking place. The influence of multimedia and participation of computing vendors in video dial tone service are part of the evolving infrastructure and service development for communications.

Many cable companies are integrating horizontally. As seen in Table 5, the top-25 cable multisystem operators (MSOs) offer cable services, but they also back or own CAPs. Moreover, they have been involved in PCS trials or license applications, or they own cellular properties. Initially, only two cable companies owned cellular properties—Comcast and Century Communications. Now, U S West by virtue of Time Warner and Southwestern Bell Corporation by virtue of Cox Cable may own cellular properties as well. In addition, Viacom could own cellular properties by virtue of the BellSouth Corporation involvement, because BellSouth is one of the largest cellular operators in the United States.

In many cases, cable companies have entered the CAP business in cities where they have cable services. Teleport is a good example of the horizontal integration and influence of the cable industry. It is jointly owned by four cable companies—Cox, TCI, Comcast, and Continental Cablevision. TCI owns Digital Direct. Comcast is 50 percent owner of Eastern Tele-Logic and 20 percent owner of Teleport. Continental has an alternative access subsidiary called Alternet, which is considering networks in several cities.

Teleport also has joint partners, as seen in **Table 6**. They are not owners, but Teleport has signed agreements with them with the goal of being the glue that holds everything together. Teleport's strategy includes a small network specifically within the CAP region and a high-speed SONET ring serving the larger contiguous circle. The concept is based on a mega head end that is the distribution center for cable service areas that are not necessarily allocated with any view toward contiguous serving areas.

Time Warner's AXS unit has four CAPs and plans to build many more with the U S West investment. It has also worked with U S West in other endeavors, notably in the United Kingdom, with TCI to deliver cable telephony. Increasingly, cable companies are delivering video dial tone as well. The cable companies are running along with AT&T video dial tone trials in Denver. TCI has trial PCS services in the U S West serving area. It has a heritage there with some of the wireless business.

Comcast is among the most horizontally integrated companies. It is the third largest cable MSO, with 2.8 million subscribers. It owns cellular properties in several contiguous states and has invested $100 million in Nextel, which has aggressive plans to upgrade its network to digital ESMR and compete with cellular service. MCI has recently purchased 17 percent of Nextel for $1.3 billion. Comcast is also involved with telephony and cable services in the United Kingdom. It owns 49 percent of Eastern Tele-Logic and 20 percent owner of Teleport. It is one of the few horizontally integrated companies that has used its cable-provided assets to offer services other than cable. For instance, it aggressively uses fiber, connecting a CAP fiber network it owns in Philadelphia to base stations to carry traffic to carrier POPs. It also provides business customers with direct PBX to cellular site connections using CAP-provided fiber networks. Thus, it is using the CAPs as big trunking networks.

Six months ago, cable was viewed as the primary competitor of RBOCs. Cable companies were pursuing regular phone service through investment in CAPs, which allowed them to serve the business markets. They were also aggressive in pursuing PCS licenses and moving into the cellular business. Now, through joint ventures and acquisitions, the evolution of local access has changed.

PCS

The dimensions of the emerging PCS market will result in the development of different markets, as illustrated in **Exhibit 2**. Many of these services will be in endeavors in which consolidation will produce some synergy. Comcast, for example, is an interesting case study of synergy across markets. Some of the new markets will be serviced over the newly licensed spectrum the FCC has allocated. The cellular carriers will be very aggressive in offering PCS-like services over their cellular spectrums. As they expand their capacity with digital, PCS providers will have difficulty in the 1996–97 timeframe, when they actually obtain the spectrum to build their networks to compete with the cellular providers. The cellular carriers will work hard to preclude that competition.

Exhibit 2. The Dimensions of the Emerging PCS Market

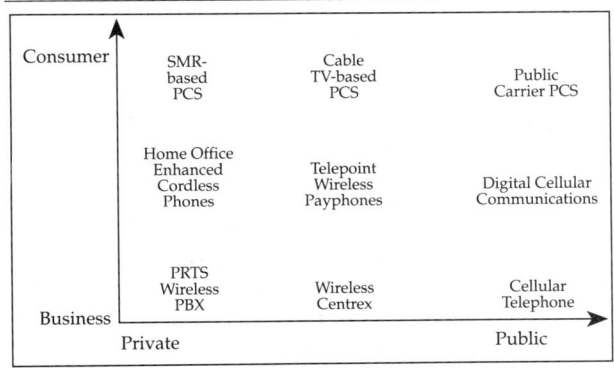

Source: Yankee Group.

PCS presents many different opportunities. **Exhibit 3** illustrates the convergence in PCS of voice, data, and even computing. In voice transmission, for example, cellular will converge through the evolution to digital cellular networks, which will expand capacity that will allow these companies to offer services for the mass market. Digital microcellular services are synonymous with the concept of PCS. From the consumer's standpoint, PCS will enhance cordless communication and provide telepoint, which is a wireless pay-phone-type service. The data-carrying element of PCS comes mainly from a paging heritage. Paging is the first and so far the most successful of personal communication services, with 15 million subscribers and growing. Other services PCS will move into include mobile data, which will transmit data calls over the cellular net-

Table 6. Teleport's Parentage and Their Influence

Cable Company	Relationship	CAP Subsidiary/Backing	Existing Franchises in Potential CAP Cities
Cablevision Industries	JV Partner		West San Fernando Valley, CA; Bell Glade, Marion County, Okeechobee, St. Augustine, and Winter Garden, FL; Fort Benning, GA; Dearborn and Wayne, MI; Saratoga Springs, NY; Philadelphia (area 2), PA; Clarksburg and Fairmont, WV
Comcast	20% owner	Eastern Telelogic	Several locations in California; Mobile, AL; Boco Raton and West Palm Beach, FL; Pontiac, MI; Philadelphia, PA; Indianapolis, IN; Trenton, NJ
Continental	20% owner	Alternet	Los Angeles, CA; Cambridge, MA; Lansing, MI; St. Paul, MN; Reno, NV; Richmond, VA
Cox	30% owner		San Diego and Santa Barbara, CA; Macon, GA; Cedar Rapids, IA; New Orleans, LA; Omaha, NE; Oklahoma City, OK; Roanoke, VA; Spokane, WA
Crown Cable	JV Partner		Alhambra and Riverside, CA; Fort Carson, CO; Bowling Green and Troy, MO; Wassau, WI
InterMedia Partners	JV Partner		Tucson, AZ; Los Gatos, Milpitas, Mountain View, and Santa Clara, CA; Fulton County, GA
Maclean Hunter Cable TV	JV Partner	Selkirk Communications	Ft. Lauderdale and Hallandale, FL; Detroit (59%), MI; East Orange and Jersey City, NJ
TCI	30% owner	Digital Direct	San Mateo, CA; Boulder and Denver, CO; Washington, DC; Daytona Beach and Miami, FL; parts of Chicago, IL; St. Louis, MO; Bloomington, IN; Reno and Carson City, NV; Buffalo, NY; Eugene, OR; Pittsburgh, PA; Sioux Falls, SD; Corpus Christi, TX; Dallas, TX; Galveston, TX; Salt Lake City, UT; Seattle-Tacoma, WA; Madison, WI
Times Mirror Cable Television	JV Partner		Phoenix, AZ; Chula Vista, CA; Springfield, IL; Amherst, MA; Providence, RI; De Rio and Midland, TX; Weirton, WV
Viacom Cable	JV Partner		Marin County and San Francisco, CA; Dayton, OH; Nashville, TN

Source: Yankee Group.

Exhibit 3. Development of the PCS Market

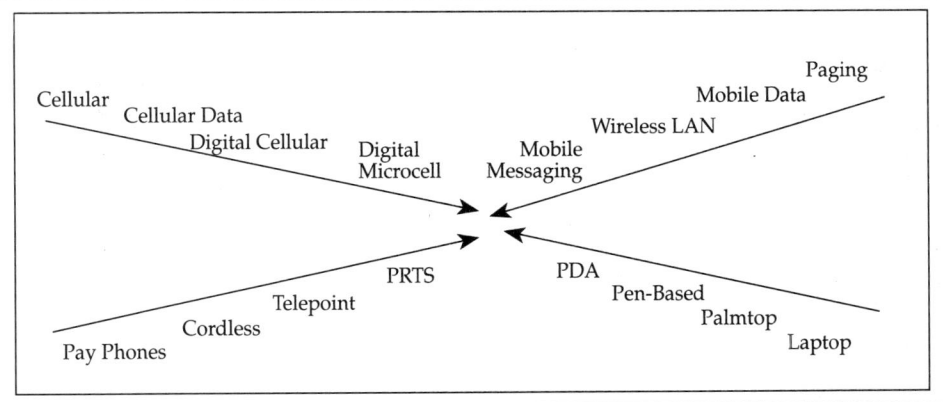

Source: Yankee Group.

Note: PRTS = Private Radio Telecommunications Services.

work when McCaw releases cellular digital packet data, a new protocol for sending data; wireless LAN; and mobile messaging, which is one of the key synergies between AT&T and McCaw—the prospect of easy linking for E-Mail and mobile messages over traditional cellular networks.

Part of PCS will be the whole computing element and mobile computing. Mobile computers with PCMCIA slots (small, credit-card-sized floppy disk drives) can start to offer all sorts of wireless communication services. Despite some disappointments in the early personal digital assistants (PDA) such as the Newton, the ultimate vision of a PDA is certainly going to be an element of PDA messaging over wireless networks. Wireless will start to have an increasing impact as second-generation PDAs with wireless communications capabilities are developed.

Question and Answer Session

Mark Lowenstein

Question: Please estimate the RBOC's true costs in revenues lost to the CAPs. Why don't the RBOCs simply buy out the CAPs?

Lowenstein: We thought they would before the cable companies became involved in the equation. Effectively, however, the RBOCs are participants by virtue of their cable company involvement. If Bell Atlantic should acquire TCI, it would also acquire TCI's CAP assets, and it is also buying CAPs outside of its service area. What will be interesting is how it deals with those companies within its service area. The gloves are off in other-region competition. An RBOC has no reason not to purchase a CAP out of its region. With cable company involvement, the RBOCs will increasingly move out of their own regions.

Question: Please explain residual interconnect charges and their significance as CAPs begin to colocate with RBOCs.

Lowenstein: To accommodate CAPs, the RBOCs must do some work in their central offices to enable the interconnection to happen, and that will be a cost element. It is one of the RBOCs' justifications for charging what they are for interconnection. They are trying to recoup some of the costs of providing interconnection. There is no blanket rule for interconnection agreements; each is negotiated on an individual basis. Some have been amicable, and others are under dispute. MFS states that Bell Atlantic is difficult to deal with and is charging the CAPs too much for interconnection. We do not know how much of that is posturing and how much is reality, but that dispute has been the most public. Teleport and NYNEX have not had the smoothest of relationships on interconnection, but it has been smoother than Bell Atlantic and MFS.

Question: How sustainable are CAPs? They provide backup or emergency service, but will RBOC improvements in reliability eventually push them out?

Lowenstein: The CAPs are facing a lot of competition. They built their first generation of business on backup, and it is only about a $200 million business. If they want to leap-frog and grow to the extent analysts have been forecasting, they must look for new types of service to provide. They cannot rely on providing backup, because the RBOCs have aggressively put in their own fiber and have very sophisticated networks now serving business customers.

NYNEX serves more than 100 buildings in New York City and 75 buildings in Boston with the same type of network that CAPs have, which is one of the reasons for CAP consolidations in the Boston market; Boston had four or five CAPs a year ago and now has only three.

The Cable Communications Industry

Barry A. Kaplan, CFA
Vice President
Goldman, Sachs & Company

> Investors value cable companies based on strong and predictable cash flows, despite modest growth and lack of earnings. Technological, regulatory, and competitive dynamics will continue to influence the industry and valuations.

Some of the fundamental issues affecting the cable industry in the past have become almost irrelevant in the face of the transactions taking place today. This presentation will focus on seven questions:

- Why do cable companies rarely have earnings?
- Why is that good?
- What is a cable company worth?
- How will regulation affect the business and valuation of cable companies?
- Why are telephone companies buying cable companies?
- What competitive threats does the cable industry face?
- What will the successful cable company of the future look like?

The answers to these questions should provide a better understanding of the dynamics of the industry.

Cable Company Earnings

Cable companies have essentially no earnings for two reasons. One is the nature of cable industry management. This industry was founded and developed by what many people refer to as a bunch of cowboys who went around the country climbing poles and stringing cable. Many of the cable operations were bootstrap. In the early days, cable was done on a financial shoestring, and entrepreneurs were constantly in danger of financial disaster. Their approach to the business was one of basic entrepreneurial interest—maximize the "took-ins" relative to the "took-outs" (the cash coming in relative to the cash going out). Such companies are run to maximize tax-sheltered cash flow, net asset value, and overall cash-on-cash returns on equity.

Another aspect of cable that affects its earnings record is its fundamental character: It has always been highly predictable. A cable entrepreneur or businessman has always been able to predict not only the direction of cash flow growth but also, within a few percentage points, how rapidly it would grow. Cable is essentially a recurring-revenue business, one that earns money from many people and is not dependent on an individual customer. It is therefore a highly predictable business with very consistent historical growth.

These characteristics make the business very leverageable, and the marketplace has provided substantial asset liquidity. For most of the industry's history, certainly for the past 10–15 years, the private market has almost been as liquid as the public market. Whenever a cable asset was for sale, investors had a good idea of what the price would be. Most of the time, the companies had a buyer waiting if they wanted to sell. The merger market multiples averaged 10 to 12 times earnings before interest, taxes, depreciation, and amortization (EBITDA), typically on a running-rate basis.

That asset liquidity obviated the companies' need to cover debt amortization from internal cash flow, which the cable industry has avoided for long periods because the companies always knew they had the ability to sell the asset or part of it if they wanted to do so. To the extent that cable companies have always been made up of multiple franchises and multiple systems, they have had discrete assets they could sell off, one piece at a time to generate cash. That situation gave them great ability to leverage their businesses. Typically, this industry is leveraged 6–7 times EBITDA. Leverage has been very attractive to the companies because it allows for true maximization of equity returns to shareholders. In that respect, the cable business is similar to the real estate business, in which leverage maximizes return.

In many other respects, however, the cable industry is very different from real estate and much less subject to the problems that the real estate industry has gone through during the past several years.

Another reason for the cable industry's lack of earnings is its capital intensity. This business has typically required about $1,000 per subscriber to build the basic plant. Even at maturity, the industry spends about $50 in capital per subscriber per year to maintain plant and perform necessary upgrades. That capital expenditure generates significant depreciation, which minimizes earnings.

The No-Earnings Advantage

The cable industry's lack of earnings frees the companies to be more flexible in running their businesses and making strategic decisions, particularly compared with how telephone companies make business and investment decisions. Because the companies do not have any earnings, they have no concern about earnings dilution in an acquisition. The cable industry has never had to consider the impact on earnings in any investment decision. This freedom has allowed the companies to make investments, particularly investments in unconsolidated affiliates and entities that would not take in any cash flow or earnings but investments that still would incur a financing cost. In a situation in which dilution to earnings was a concern in purchasing assets (beyond the dilution that companies would incur by buying 100 percent of an asset and expecting at least to get the cash flow and the earnings from that asset), they would be free to buy pieces of assets that would add nothing to their profit and loss except the interest expense from the debt they incurred. They could make investments simply based on the belief that they would add shareholder value.

Lack of earnings has also provided tax shelter. This industry has always had an aversion to paying taxes. John Malone, chief executive of Tele-Communications (TCI), once said, "It is time for me to retire when TCI starts paying taxes." Tax avoidance has been an important aspect of the whole strategy—as much leverage as can prudently be overlaid onto the company. The companies can maximize their equity returns to the extent that their returns are higher than their debt costs.

Valuation

The valuation method for cable is relatively straightforward. Consider the return analysis for an unlevered cable system as shown in **Table 1**. This cable company starts out with 100,000 subscribers and monthly revenue per subscriber of $30. Total annualized revenues are $36 million. The operating cash flow (EBITDA) is $17 million based on a 47 percent operating cash flow margin, which is typical for a cable company today. Capital expenditures are $50 per subscriber per year, or $5 million. This completely unlevered company initially has no cash and no debt and is valued at 10 times cash flow with enterprise and equity values of $170 million.

This utility growth model reflects the mode the industry has been in for the past several years. This model assumes the cable industry will continue its growth on a quasi-utility basis. By assumption, no dramatic technological changes will affect revenues generated or money spent on capital. Assumptions are based on what the industry has been doing recently—3 percent subscriber growth, rate increases at the pace of inflation of 3 percent, some modest increase in pay-per-view penetration and buy rates, and about 20 percent annual growth in advertising revenues. The operating cost growth is somewhat higher than inflation, and programming costs are assumed to increase about 15 percent a year.

Projecting out five years, the subscribers increase to 116,000. Monthly revenue per subscriber is $38,

Table 1. Return Analysis for an Unlevered Cable System

Item	Year 0	Year 5	Annual Growth Rate
Subscribers	100,000	116,000	3%
Monthly revenue per subscriber	$30	$38	5
Revenues ($millions)	36	53	8
Operating cash flow (OCF) ($millions)	17	26.5	9
Margin	47%	50%	
Capital expenditures at $50 per subscriber per year ($millions)	$5.0	$5.8	
Net cash (3% interest assumed) ($millions)	0	77.7	
			Internal Rate of Return
Equity value: at 10 × OCF ($millions)	$170	$342	15%
Equity value: at 8 × OCF ($millions)		290	11

Source: Goldman, Sachs and Co.

up 5 percent a year because of a combination of rate increases, additional services such as pay-per-view, and the allocated portion of advertising revenues that the cable company generates spread over the subscriber base. Total revenues are $53 million, based on a compound annual growth rate of 8 percent. Operating cash flow, which is growing 9 percent a year, is $26.5 million. The margin has gradually moved up to 50 percent, which is typical when looking at the prospective costs of the industry on this amount of revenue growth. The capital expenditures are the same at $50 per subscriber per year, growing to a total of $5.8 million. The company is generating $77.7 million in free cash flow. Assuming the same 10 times cash flow multiple as at the beginning of the period, the company is now worth about $342 million. Assuming that this business's growth is slowing modestly, an analyst might want to use a lower multiple. So, at 8 times cash flow, the value is $290 million and the internal rate of return (IRR) is 11 percent, which is okay but not compelling.

Leverage makes an enormous impact on the return analysis. Typical cable companies are leveraged at a level of 6 to 7 times cash flow. The model in **Table 2** assumes a multiple of 8 to emphasize the impact of leverage, and in fact, many cable companies are leveraged 8 times cash flow or more. The operating parameters are the same as in Table 1, except that this firm began with net debt of $135 million. Thus, its starting net equity value is only $35 million. The operating parameters in Year 5 are exactly the same, but the company used its free cash flow to pay down debt, so it now has only $80 million of debt. It has $185 million equity value in Year 5 at 10 times cash flow; even at 8 times cash flow, its equity value is $131 million. At 10 times cash flow, the IRR is 39 percent as opposed to 15 percent for the unleveraged company; at 8 times cash flow, it is 30 percent compared with 11 percent.

As seen in this valuation exercise, using leverage, some very compelling equity rates of return can be generated on modest growth. The cable industry has played the high leverage game for more than ten years to generate substantial equity returns for investors, even though the growth of operating cash flow has slowed to about 9 percent. Historically, industry growth has probably been in the low to mid-teens, but going forward, it will be more like that 9 percent, assuming no new technological developments occur. In fact, however, the growth rate will probably decline during the next 12–18 months because of rate reregulation.

Leverage is a simple concept that could apply to any business. It is nothing unique to cable TV, but in fact, the cable industry has always been unique because of the predictability of its cash flow. In the late 1980s during the leveraged buyout craze, many LBO practitioners tried to achieve high returns on low cash flow growth with other businesses such as retailing, consumer products, and manufacturing. In many cases, they got into trouble and the deal failed. The failure was based on the fact that, although they could make the spreadsheet say whatever they wanted, the bottom line was that the business was not nearly as predictable a cash flow generator as the cable TV business. In no year in the history of cable has the cash flow not been significantly higher than the year before. This obviously may change under regulation.

Analysts and investors value cable companies based on cash flow, placing a multiple thereon. Assume, as in Table 2, that in five years the cable company is worth 10 times the operating cash flow and has a value of $184.5 million. Discounting that value at 20 percent a year (our assumed investors' required rate of return), an investor would be willing to pay $74.5 million for that cash flow today plus $135 million of starting debt, or $209.5 million. In this exam-

Table 2. Impact of Leverage on Cable System Internal Rate of Return
(assumes 8 times debt to cash flow)

Item	Year 0	Year 5	Annual Growth Rate
Subscribers	100,000	116,000	3%
Monthly revenue per subscriber	$30	$38	5
Revenues ($millions)	36	53	8
Operating cash flow (OCF) ($millions)	17	26.5	9
Capital expenditures at $50 per subscriber per year ($millions)	5.0	5.8	
Net debt at 5% (cost of debt) ($millions)	–135	–80.5	
			Internal Rate of Return
Equity value at 10 × OCF ($millions)	$35	$184.5	39%
Equity value at 8 × OCF ($millions)		131.5	30

Source: Goldman, Sachs and Co.

ple, an investor would pay a 12.3 multiple of current cash flow ($209.5 million in value divided by $17 million in current operating cash flow). That is how a business with modest growth achieves a valuation with a relatively high cash flow multiple.

Using a terminal multiple of 8, which is below what most people would expect in the cable business, and even excluding any technological developments, the current value is 11 times current cash flow.

The Effects of Regulation

In many respects, regulation has come full circle in the cable TV business. Prior to 1986, the industry's basic rates were heavily regulated by the local cities and towns granting cable franchises. To get a rate increase, a cable company had to go to the town fathers and plead for it. Rate increases were very difficult to obtain, and rates for basic service typically rose more slowly than the inflation rate.

To compensate for the inability to raise basic rates, the industry raised pay TV rates rapidly. As a result, it went through a period of rapid growth in basic subscribers but a stagnation of pay TV because the price-to-value relationship of pay versus basic was out of sync. In fact, one prominent cable operator confided to me once that the problem with pay TV is that it is trying to sell a $5 service for $10. In many respects, that continues to be the case, although the pricing of pay services has been virtually stagnant since 1986 or 1987.

The cable communications policy of 1984 did not have much effect on the industry until the beginning of 1986. At that time, basic rates were deregulated and cable operators could charge whatever they wanted. The period from 1986 through 1992 was one of rapid growth in basic rates, substantially in excess of the inflation rate, but this rate of increase gradually diminished. Now, even in the absence of regulation, rates would probably not rise faster than inflation. Pay TV rates also flattened because the industry tried to improve the price-to-value relationship.

The incremental costs associated with raising basic rates were low, so the industry saw a dramatic acceleration of capital. Cash flow increased 15–20 percent a year, particularly during the 1986–88 period. The flow-through of the rate increases to the bottom line was 80 or 90 percent. Commensurate with that increase in profitability was a substantial jump in cable system values. Merger market prices had been at the traditional 10 times cash flow multiple, but the multiple began to move beyond 12 times. In fact, some deals were valued at 14, 15, and 16 times cash flow as buyers saw a never-ending ability to raise rates and finance very high percentages of the purchase price with relatively low-cost debt. The merger market also went through a frenzy in terms of volume of activity in the trading and retrading of systems. Some systems were bought and sold three or four or five times in the course of a few years. The financial players were very active in this market as were the traditional cable operators.

In 1992, the party came to an end. Since the day cable rates were deregulated, unsuccessful efforts were made to reregulate them. In 1992, however, Congress passed the Cable Consumer Protection Act. On the surface, the Act appeared to be extremely onerous. Basic rates were frozen through September 1993. That freeze has been extended for those operators that have not implemented the required benchmark rate changes. At a cost-to-service hearing, cable operators could claim that their systems were operating under special circumstances so they should not be subject to the FCC's basic rate benchmarks. Their rates are frozen until their cases are resolved.

In 1993, most of the industry underwent basic rate rollbacks, which were intended to reduce consumers' cable bills. The FCC and Congress attempted to roll the average basic rate back by 10 percent, but the reality was not quite so bad. Another key provision was access to programming. This provision mandated that programming entities (cable program networks) controlled by large cable operators could not unreasonably deny cable competitors access to that programming. They also could not discriminate on the basis of their prices to competitors for reasons other than volume discounts that are directly associated with costs.

Another perceived important provision of the bill was the retransmission consent provision, which allowed broadcasters to negotiate for payment in exchange for carriage on a cable system. Although a huge fight developed between cable operators and TV broadcasters, the practical impact was essentially zero. Almost every broadcaster in the United States ended up capitulating to the system operators and allowing cable operators free carriage for their television signals.

Once all the numbers were calculated and the rollbacks were completed, the net impact was not as severe as many had thought it would be. Many thought the negative revenue impacts would be 5 percent or more and the negative cash flow impacts would be at least 10 percent year over year. For most of the industry, the average impact will be flat cash flow growth, although increasingly some of them come up with new ways to circumvent the regulation and reduce the impact. Many of the larger companies will have cash flow increases in 1994 despite regulation.

The net impact of program access is to reduce

business risk to competitors, which traditionally are wireless cable and direct broadcast satellite (DBS). Historically, wireless had difficulty getting programming access, and it had difficulty selling video services without programming. That situation has changed radically, and now almost anybody who wants to buy programming can. This access also applies to the telephone companies, which cannot be reasonably denied programming.

Investors might have expected this ruling to have a material negative impact on the value of cable systems in the merger market, and for a very brief period, it did. Because of everything else that has been happening in the industry, however, the negative results have been completely overwhelmed. Cable valuations have increased significantly in recently announced merger transactions.

Telephone Company Cable Acquisitions

As a result of rapid technological developments in the computer software and hardware, consumer electronics, cable, and telecommunications industries, a true revolution in the delivery of entertainment, information, transactional, and telecommunications service may be at hand.[1] Technology is driving this convergence, or consolidation, in the industry. Acquisition activity is far from an anomaly; in fact, the extent to which it does not happen will become the anomaly.

The proposed Bell Atlantic Corporation–TCI transaction has been the most publicized, but several others have already occurred. The acquisition movement began in early 1993 with Southwestern Bell acquiring the Hauser properties in Maryland and Virginia for $650 million. Then followed the U S West–Time Warner entertainment transaction and NYNEX Corporation's investment in Viacom. The same day the proposed Bell Atlantic–TCI deal was announced, BellSouth Corporation announced it was making an investment in Prime Cable. There has been widely reported speculation about Cablevision Systems being for sale with U S West and others as possible buyers and about Southwestern Bell Corporation and Cox Cable merging.

The numbers presented in **Table 3** are related to what is driving the business. Some observers believe that technology eventually will allow the cable and telecommunications industries to access all these markets for various services, which are already quite substantial. The mail order business is worth $50 billion, and at the same time, the video home shopping business is developing. The Home Shopping

---

[1] "Communacopia," Goldman, Sachs & Co. Research, July 1992.

Table 3. Potential Cable Industry Markets
(billions of dollars)

Shopping and financial	
Mail order	$ 50
Brokerage revenues	70
Coupons	13
Air travel commissions	5
Cable home shopping	2
Total	$140
Telecommunications	
Plain old telephone service	$ 39
Long distance	67
Local access	28
Value added network services	7
Cellular and PCS	7
900#	2
Total	$150
Information services	
Classified advertisements	$ 12
Electronic information	10
Yellow pages	9
Software	4
Broadcast advertising	30
Total	$ 65
Entertainment	
Home video	$ 12
Recorded music	9
Box office	5
Video games	5
Total	$ 31

Source: Barry A. Kaplan, CFA, based on Time Warner and Goldman, Sachs Company estimates.

Network and QVC Network already have a $2 billion market, but it will grow much larger. Other examples of markets with large potential include brokerage at $70 billion and coupons at $13 billion.

The telephone industry is at risk. For some years, the cable industry has been salivating over the telecommunications market, and that is a very big part of what this whole convergence process is about. The other information- and advertising-related services are also good markets. Home video, which will allow consumers directly to access a movie over their cable systems, is a $12 billion market, and it is a business that would be dead-on in the sights of cable operators. In the first quarter of 1994, the Sega channel, which will distribute games to users over cable, will be launched. This market will be worth $5 billion.

These new markets present opportunities for the cable industry and sources of concern for the telephone industry. Even if both networks were on an equal footing in terms of ultimately creating a full-service broadband interactive network, and each could capture 10 percent of the other's business, the cable industry would be in a better position. Today, the cable industry has 57 million subscribers, who pay an average of $30 in monthly revenues for a total of $22 billion in annual revenues. The local telephone industry has 140 million lines with an average monthly revenue of $53 for annual revenues of $89 billion. That trade-off favors the cable industry, and it provides another reason for telephone companies to own cable.

In this race to deploy full-service broadband networks and compete for new services, the telephone industry faces some regulatory obstacles that the cable industry does not. The cable industry has had its own regulatory problems, but they do not include issues of cross-ownership and the inability of the telephone companies to offer video programs in their own service areas. So far, this incapability has been successfully challenged by Bell Atlantic, although the final word has not yet been written.

Telephone companies are also subject to state public utility commission regulation of capital deployment. Thus, the state determines how capital is deployed in the telephone industry. The purpose is to avoid a situation in which telephone companies force their rate payers to subsidize or pay for their forays into new, high-risk business ventures.

From a physical plant standpoint, the cable industry is in a better position to get to the broadband and interact with the full-service network much more quickly and for much less money than the telephone industry. The network is designed with coaxial cable running into users' homes (coaxial drop). The telephone industry drop is copper and is handicapped by what I call the 80/20 rule. To get to the same place, the cable industry must replace only 20 percent of its plant with fiber, specifically the trunks and some electronics. Although in most cases the telephone industry has fiber in the trunks, because of its copper drop, closer to 80 percent of the plant mileage of the network must be replaced. Ignoring the money factor, the construction time alone to replace all that with fiber is formidable.

The cost for a telephone company to upgrade to broadband is between $2,000 and $3,000 per subscriber for a full-service network as opposed to $800 to $1,000 for a cable company. Cable operators can offer telephone service more easily than a full-service broadband network. For example, technologies available from companies such as First Pacific Networks have fully distributed switching capability that offers full telephone service at between $400 and $500 per subscriber. Plain old telephones are not a broadband service, but they are interactive. With fully distributed switching, however, that capability can be achieved without having a centralized switch.

To the extent that cable operators provide telephone service, they will link up with alternative access carriers such as Metropolitan Fiber Systems (MFS) or Teleport or even the long-distance companies that will share switch facilities. The time required to upgrade the network to true full service is two to five years for the cable industry, depending on how much of the country is completed. The telephone industry would require about five to eight years or even longer to cover a substantial portion of the country.

Telephone companies are investing in cable for several reasons. There is an explosion of new potential services. The telephone companies face a number of regulatory and physical obstacles to offering those enhanced services as part of the telephone service. Cable companies offer a significant competitive threat to their basic business, which they would like to offset by competing outside their own telephone service area. If competitors are allowed to enter the local telephone business, the telephone companies would like to be the new competitors in areas outside their service region. Doing so allows a telephone company to expand its geographic reach. For the most part, the RBOCs have been confined to their own service areas to the extent that telephone has been their principal business. Acquiring cable companies allows them to expand beyond that. To the extent that programming software assets come with these acquisitions, the union of the two gets them into the content business, which many people think has a higher margin than distribution alone.

Other Competitive Threats

One threat to the cable industry is high-powered direct broadcast satellite (DBS). High-powered DBS will dominate the rural markets, where it is not competing directly with cable. Hughes Communications will launch a satellite in late 1993 and a digitally compressed 150-channel video service early in 1994. It will be in a position to offer multichannel pay-per-view TV (40 or 50 channels) sooner than the cable industry. It will offer some traditional cable services and some specialized video program services that do not compete with cable.

High-powered DBS has some limitations in competing head-to-head with cable. The DBS technology, as it stands today, lacks local or regional capability. By definition, satellite is a nationwide beam, which cannot retransmit local broadcast signals. With DBS—a dish instead of cable—a viewer would have to switch from the dish back to over-the-air to get a local broadcast signal. There is some discussion and technology development to create local satellite beams, which would make the service much more compelling, but we have not yet seen demonstrations of that capability. True interactivity is also more difficult to achieve with satellite than with wire. We think DBS will ultimately be a successful business, however, particularly in rural and other niche areas.

Another competitor fighting the war against cable in many markets is wireless cable (MMDS). Many of these small wireless cable companies have had public offerings, and the stocks have done extremely well. MMDS companies operate at the low end of the market. They have a low-cost basic package and relatively low-cost infrastructure, assuming that they can get it loaded with a reasonable number of subscribers. Their channel capacity is limited, however. Very few of them have been able to aggregate more than 33 channels, and most have fewer than 33 channels. Also, they have only limited interactivity. Wireless cable is a line-of-sight technology, so buildings and mountains will block and limit the reach of the signal. As a result, MMDS will be a successful niche player at the fringe of the video market but never a business that approaches the size of the cable industry.

Future Success in Cable

Cable is a business in which size is increasingly becoming imperative, and the definition of "critical mass" is becoming larger and larger. In the past, 1 million subscribers was considered the critical mass, but now it is more on the order of 2 million, and some people have suggested 4 or 5 million subscribers as the minimum.

In a business with rapidly moving technology, having a say in how the technology evolves is important. A large cable operator, such as TCI, will have an enormous say in the evolution of the technology, so it can help guide development in a direction most advantageous to it. Also, the largest cable companies can participate as equity investors in that technology, as TCI is doing. It has invested small amounts of money in several technology companies. When decision time comes, it can approach these companies and say it likes their technology and would like to deploy it across its 10 million subscribers—but it would also like to own 25 percent of the company. And it may successfully negotiate a deal.

Large economies of scale will be possible in purchases of equipment and programming, both of which will become increasingly important. A 500-channel system requires a lot of programming. A lot of money must be spent to upgrade the network. The companies will need as much economy in equipment purchases as they can get.

Analysts are currently seeing a bifurcation between large and small cable companies in access to capital. A lot of capital will be needed to upgrade the network, and the ability to finance it is critical. The larger companies get much better availability and terms than the smaller ones. This differential is part of what is compelling the consolidation process in the industry. On the buy side, telephone companies will line up with the largest cable operators. Smaller cable operators that do not feel financially, technologically, or managerially equipped to deal with industry changes and evolutions will be eager to sell.

Geographic clustering is critical going forward, because advertising will be an increasingly important source of revenue for the industry. The ability to offer an advertiser a complete ADI (metropolitan area) in coverage is important. That may not be required in order to be an economic business, but it helps if the company owns all the cable subscribers in a local market. Having all its subscribers in one place offers some operating efficiencies in delivering telephone service. To the extent that the cable industry is in the telephone business, the most profitable call is one that a company both originates and terminates, so wide coverage of a local area helps considerably in that respect.

Cable operators will become more customer-service oriented. In large part, this development is being driven by competitive forces. Also, the cable industry gains from acquisitions by local telephone companies, which are typically associated with good customer service. Just the fact that the environment will be more competitive will force good customer service beyond any regulatory requirement to do so.

The cable industry will continue to build net asset value through growth and tax-sheltered cash flow. The cable companies will play the same game they have always played. Regulation will cost the industry a year or two of growth but not fundamentally alter the economics of the business. Technology will pose some increased risk because of the uncertainty of how it will evolve and what the consumer demand for it will be. Consolidation will continue and accelerate both within the cable industry and with respect to telephone companies. The net result of all of these developments will be continuing consolidating transactions, and ultimately higher asset values and stock prices.

Question and Answer Session

Barry A. Kaplan, CFA

Question: You note that cash flows seem relatively predictable for cable. What is your view concerning costs, and what do you think the growing competition will do to the cost structure?

Kaplan: The trend in costs depends on how you define costs and how you assume the world will evolve. As configured today, costs are fairly predictable, with the possible exception of programming costs. To the extent that technological developments are uncertain, capital costs become more uncertain. The question is not so much where a company wants to get in the technology development of its network but how fast it wants to get there. That question raises the issue of capital costs. Do they build a state-of-the-art network as quickly as possible and hope somebody shows up? Or, do they convert it incrementally, spend some capital, add some services, and see what the response is? The uncertainties are much greater on the capital side of the business than on the operating side.

Question: Please elaborate on the significance of some of the newer technologies such as DBS and satellite master antenna and some of the difficulties that are on the horizon for DBS.

Kaplan: DBS will be digitally compressed and have 150 channels and will be a good service for people who do not have access to cable. It will be a compelling program service that will probably offer pay-per-view a year or two sooner than cable. One of its overwhelming limitations is that it lacks local transmission of TV signals. If viewers cannot get local television stations from the service, DBS will be difficult to sell. In larger, less rural areas, it must compete with a large installed base of cable. Inertia alone makes cable difficult to displace. It does have an advantage of being largely a fixed-cost business, although it requires lease of satellite capacity, which is the largest cost. Once the companies cross the breakeven line, however, DBS could become incrementally quite profitable. The operators can amortize that satellite over the whole country, which is a big potential customer base. They do not have to get high penetration to make the business work.

Question: We have a vast array of potential services and channels. What is the evidence that customers will value that capability highly?

Kaplan: Many of the businesses and services the cable industry is considering providing are not necessarily new services. They are established businesses that would be done somewhat differently through cable. For example, the video-on-demand business is accomplished today by people driving to the video store and renting videos. It is an established $12 billion business. The telephone business is an established business that is in demand and that people use. Whether people will give up their existing telephone service provider is a pricing and service issue more than a demand issue. Whether some of the other, more esoteric services will be in demand or not is unclear, but with so many of them in prospect, inevitably some of them will stick and many will fail.

Question: What will happen to pay-per-view and video rental companies like Blockbuster Entertainment?

Kaplan: Pay-per-view has a couple of problems to solve. It may not need true video on demand, but at a minimum, it needs near video on demand, which is a wide choice of movies. The critical driver in pay-per-view is choice. Buy rates will not be very high for a choice of two to four movies, especially if any one of the given movies begins only every two hours. It must offer more of an on-demand service in which viewers can go into their houses and be within ten minutes of any of the top 30 or 50 movies they could rent in their video stores. All the tests so far show a directly proportionate relationship between buy rates and the amount of choice.

Another factor that would help pay-per-view is the improvement of release windows for movies. Currently, the lag between when a movie comes on tape and when it is available on pay-per-view is six to eight weeks. To the extent pay-per-view grows and becomes a larger revenue opportunity for the studios, these windows will gradually improve.

Pay-per-view service will hurt the video rental marketplace. I do not want to comment specifically on Blockbuster. It is a unique company that has been taking very aggressive and substantial steps in acquisitions, joint ventures, and new businesses to diversify and take advantage of the distribution capability it has through all its stores. The traditional video rental business will not be a great growth business in the late 1990s, however.

Question: In "Communacopia," the argument was made that cable beats telcos for delivery of multimedia. Would you revise that position in light of recent market activities?

Kaplan: On the contrary, we think everything that has happened recently is an affirmation of the original thesis of the report and a recognition by a substantial part of the telephone industry that the cable infrastructure has enormous value that is worth paying large amounts of money to acquire. Part of that attitude is defensively offensive, however. If somebody in an industry represents a potential threat to a basic business, that business will try to buy it out.

Question: Cable has the big pipe but what about the switch?

Kaplan: The switch is much less important than the pipe because, over time, the switch itself will become a commodity and not that hard to install. Access to switches is easily available. A company does not have to own the switch; it can use MFS's, Teleport's, the local telephone company's, or AT&T's switch. Installing a switch does not take that long and is not that expensive compared with laying fiber.

Question: If convergence will require higher capital spending by cable companies and result in a dislocation of EBITDA and free cash flows, will cash flow multiples no longer be used as heavily for valuation purposes?

Kaplan: EBITDA multiples will continue to be used, but the multiple investors are willing to pay on that EBITDA may decline to the extent the business is perceived as more capital intensive than it used to be.

Question: You have talked about cash flow predictability and focused on revenues. Are costs as predictable? How will revenue predictability be affected by satellite broadcast?

Kaplan: Costs are generally quite predictable and in many cases such as programming have contracted. Revenues will become less predictable over time, not just because of competition, but because of regulation and the blue sky nature of new services.

Question: Given access legislation, why is everyone trying so hard to secure programmers?

Kaplan: Obviously, if distribution becomes more competitive, demand overall for programming increases and the value of high-quality programming will rise.

Question: What is your impression about 28-GHz wireless cable?

Kaplan: If it does everything its proponents claim, it could be quite a competitive technology. It is really unproven as yet, however, and I think the capital costs are higher than we have been led to believe.

Question: Some observers have viewed the convergence of telcos and CATV as an exit strategy for cable players. What is your point of view?

Kaplan: I believe there is some truth to that view. Communications is becoming a business for very big companies. If you cannot or do not want to get much bigger, selling to a telco is an attractive alternative.

Question: Are there any further regulatory or legislative repercussions because rate freezes and accompanying regulation goals appear to have backfired in many areas?

Kaplan: Yes, there is a real risk that the FCC will come back and take another whack at the cable industry. This eventuality would increase the pain for the industry and ultimately force it to seek redress in the courts.

Question: Will RBOCs win the right, either through the courts or Congress, to offer cable in existing service areas?

Kaplan: Ultimately yes, but only as a competitor to cable, not as a buyer of the cable system. The regulatory forces are inexorably moving toward promoting competition in all areas of communications.

Question: What is involved in upgrading the telco plant to broadband (i.e., what is in the $2,000 to $3,000 per subscriber upgrade cost)?

Kaplan: Most of the cost is in laying fiber and/or coax in the local loop. Upgrading will require thousands of miles of plant, plus optical converters and other electronics, and ultimately ATM switching and video servers.

Question: We hear about the necessity of scale as a justification for some mergers. With all providers assured equal access to programming, what is the disadvantage of being a smaller cable system?

Kaplan: Equal access does not mean equal price. Moreover, there are significant economies of scale in the delivery of telephone and other services. In addition, the increasing capital intensity of the business requires access to big capital.

Question: Telco's one-year capital expenditures equal the total spent by the CATV industry. So, can telco duplicate CATV capabilities with one year's worth of capital spending?

Kaplan: No, the telephone industry must do and spend much more just to get to the same place as cable.

Question: Do you see two broadband wires going into each home in the near-to-intermediate timeframe? If so, what happens to profitability/value (cash flow)?

Kaplan: Ultimately, I believe two wires will go into the home in many places, especially urban areas. In rural areas, the economics may not justify two wires. Clearly, governmental policy is to promote two wires, and both the telephone and the cable industries seem willing to do it. As a result, margins in both the telephone and cable business may decline, but the $64,000 question is whether the new revenue sources will offset the impact of competition.

Question: Please explain asynchronous digital subscriber line (ADSL) technology. Is it a threat to cable's technological advantage?

Kaplan: ADSL is simply a compression of technology that allows limited transmission of video over copper twisted pair. I do not know anyone in the telephone industry who truly believes that ADSL is a long-term viable alternative to true broadband capability. It is an interim technology.

Question: Is there any way to quantify the economics of overlaying PCS on the cable infrastructure?

Kaplan: In my opinion, the economic benefits of the PCS overlay over cable have been exaggerated. In principle, however, the cable network could provide a cheaper means of transport of voice traffic between call sites and the switch from "satellite" cells to a hub cell at a much lower cost than leasing lines from the telephone company. This approach might also permit concentration of processing intelligence in the PCS hub cell site and thus reduce the capital cost of the "satellite" cells.

Question: As customers are given more control over the programming they select and watch, what is the role of advertising and what will it be in the future?

Kaplan: That is a very interesting question. The future will pose great new challenges to advertisers in trying to reach their target audiences. Broadcast television and cable will always be available, although to slightly shrinking audiences. In the on-demand environment, however, advertisers must produce advertising that is either useful enough or entertaining enough that consumers will choose to view it.

Question: Have any effective studies determined the demand for 500+ channels, interactive games, and other multimedia services?

Kaplan: Probably a dozen or more market tests are going on as we speak, none of which has as yet provided any definitive results. In the end, this capital deployment inevitably will have some "Field of Dreams" component.

Question: If you assume that Bell Atlantic will acquire but never sell TCI, why are you comfortable with a terminal operating a cash flow multiple of 8–10 times as opposed to a free cash number?

Kaplan: The cash flow multiple is simply a shorthand way of capitalizing a longer term stream of free cash flow. In the end, if you embody some assumption about what percentage of EBITDA ends up as free cash flow, then both methods amount to the same thing.

Question: Could investors interpret prices telcos are paying for cable companies more as a sign of telco desperation than a sign of attractiveness of cable as a stand-alone business? What is the implication for stock investors?

Kaplan: I think the moves are both offensive and defensive. Telcos want to offset the negative impact of cable offering telephone service in their service territories and also want to expand their geographic sphere of influence in a networking business that is not totally dissimilar from what they do now.

Question: Most cable markets are extremely fragmented. Even the largest cable companies have few large concentrations of subscribers—isn't there a risk that cable companies will overpay as they try to consolidate in order to reap economies?

Kaplan: Overpayment is always a risk, but the current state of the capital markets will impose some discipline. In addition, you may begin to see more asset swaps to rationalize properties geographically, and in these instances, only the relative value will be important.

Question: Which companies stand to benefit from satellite broadcasting beginning next year?

Kaplan: All the major programming companies will be selling to DBS: Turner, Viacom, Time Warner, and so forth. The most well-known service, DirecTv, is owned by GM Hughes.

Question: What is the likelihood of getting C-band (free) satellite dish users to convert to DBS (pay) in rural areas?

Kaplan: They are highly likely to convert to the extent that all the major satellite services will be increasingly encrypted or scrambled so C-band subscribers will have to pay for the service.

Question: When should cable companies be valued on earnings, not cash flow? When does a market mature?

Kaplan: As long as the industry manages itself to maximize tax-sheltered free cash flow, it will never be a big earnings generator. The lack of earnings is less a function of the maturity of the industry than it is a function of the capitalization.

Strategic Issues in Multimedia: AT&T's Perspective

Richard S. Bodman
Senior Vice President
AT&T

> No one knows exactly how the multimedia landscape will develop, but companies are betting millions of dollars on the outcome. Some companies are competing to control content, others are competing to control distribution, and still others are competing to control the integration of the two.

Rapid changes in the telecommunications industry have led companies to adopt a variety of far-reaching strategies, many of which include multimedia. This presentation reveals how AT&T has structured itself to deal with the changes and examines the current competitive situation in multimedia, as well as its potential impact on the future of AT&T.

AT&T's Strategy

No single company has the necessary capital base available to build extensive global plant and equipment, so AT&T has a very simple strategic backdrop.

Global Networks

Our job is to attract customers by extending the capability of networks around the globe. Although AT&T does not necessarily own the network wires or wireless equipment, it has the best telecommunications networks in the world. The equipment may belong to someone else, but AT&T administers the network to provide customers around the globe with telecommunications solutions.

Another important part of AT&T's business is supplying systems and hardware for networks. We build networks anywhere in the world, not just for ourselves but for anyone interested in buying them. Although two-thirds of our revenues are based in communication services, supplying systems and hardware for networks contributes $13 billion. Worldwide, companies and governments are spending hundreds of billions of dollars to build communications systems, and they will spend more during the next five or six years.

Products & Systems

AT&T also has product businesses. It sells access terminals—what some people call telephones. Few people use ordinary telephones any more. Most telephones are computers; they are devices that access networks. They have storage and forwarding capability and video screens. They do not look like the telephones of yesterday. Almost every service in the telecommunications business, whether accessing a voice mailbox, talking in a video telephony mode, or setting up a conference call in the future, will require a different kind of device so users can manage transactions or other events easier than by telephone.

The important task is to marry those devices with the new services as we offer them in the market. For example, McCaw Cellular Communications initially offered a subsidized piece of hardware so it could launch cellular telephony. This model is practical for most future applications. AT&T's stake in these access terminals is very important. We want to unite them with the service part of our company so that, as we introduce services, the hardware to use them is available.

Systems Integration

Telephony companies have to talk to customers in the customer's language. Some of our smaller customers want more than a toolbox. They want us to help them run their businesses better. Some want AT&T to provide only the necessary tools and services, while others want us to take over certain tasks for them.

AT&T feels enormous pressure from customers to broaden our role of integrating the network and hardware capability, including our computing capability, to provide better solutions for them. In this regard, AT&T differs from the rest of the market in two ways.

First, it is the only company that unites the idea of selling the service, the software, the systems that deliver the service, and the access hardware customers need to use the service. If that packaging is done well, it has a strategic advantage over its competitors.

Second, AT&T is one of a few companies that examines the fundamental processes to determine what can be done better. Many competitors claim they do not need to do any research and development. They believe they will never fall behind, because they will quickly buy what is available in the market. The result, however, is an endless race in which all competitors reduce costs and catch up with the leader. As evident in the computing industry, few companies have room to make money when the product is a commodity. Eventually, somebody will figure out how to do it better and where all the value will be. We hope to be one of the few companies that have value in proprietary products and services, giving us an advantage over our competition.

AT&T's Structure

AT&T has four major operating businesses, as seen in **Exhibit 1**. One is *communications products*. Think of a set of access terminals that look less like today's telephone and more like computers or television sets. About $45 billion of our business is in *communications services*. We also provide *network computing*, which is why we acquired NCR Corporation. The world of telecommunications is computer driven; it is no longer just wires, fibers, or ether. Computers handle how companies switch, how they set up packets, how they carry traffic, and where data bases are held. They handle the billing and the complicated problem

Exhibit 1. AT&T's Operating Businesses

Communications Products	Networked Computers
Communications Services	Network Systems

Shared Advantages Platform	
• Our Networks	• Access Relationships
• Brands	• Bell Labs/Shared Technology
• Customer Base	• Financial Services

Source: AT&T.

of converting protocols. The fourth business is *network systems*. As expected in any large company, the edges of these pieces are rough and do not fit together as neatly as in the exhibit. The trick is to make them fit better.

At the base of all these businesses is a shared-advantages platform. We encourage each of our business units to take advantage of those common factors—the networks, the brands, and the enormous customer base—so that we all make money.

Although we do not discuss our Bell Labs technology and our financial services as much as we should, they are an important underpinning of our global strategy. Calling cards, US Direct, World Connect, and financing for construction of plant and equipment overseas are important common platforms that give us strength.

The Multimedia Landscape

During the 1980s, two significant trends appeared that would alter the future of telecommunications. The first trend was the increased use of computers. A few years ago, AT&T realized computing was critical to our company's success. At the time, our scientists were discussing problems in technical telecom terms, while the computer industry worked on the same problem using technical computer terms. Even though we had our own computing business within AT&T, tensions and misunderstandings occurred. We decided that the only way to deal with this problem was to bring a large computing business, at the front edge of technological change, into the organization and force employees to learn each other's language so AT&T could take advantage of the fact that networks are computers, not wires. That business, as you know, was NCR.

The second trend began in 1988: Many people were using cellular telephones to make calls from their cars. They were the wealthiest and most adventuresome of our customers. Using a car phone then was not without problems, however. Customers needed a phone that was voice activated and that resisted scanning and eavesdropping. It was a terrific opportunity to build an industry, but AT&T was not in the cellular business until we formed an agreement with McCaw. Starting in 1990, the cellular business became a major thrust for the company. The acquisition of McCaw will allow us to begin delivering this new service to the world.

AT&T's vision of the multimedia conceptual landscape is based on the relationship between content players, distribution companies, and consumers, as depicted in **Exhibit 2**. Content players provide entertainment, educational and training programs, games, information systems, software distribution,

Exhibit 2. The Multimedia Conceptual Landscape

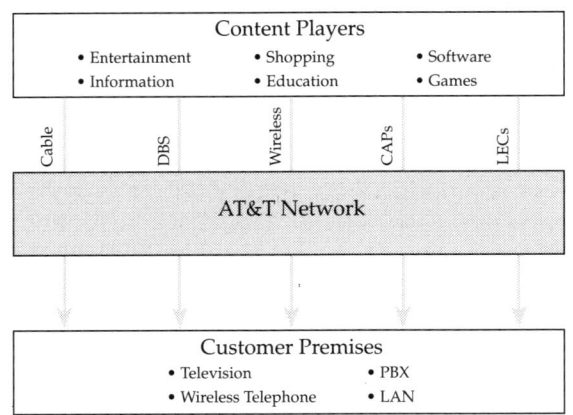

Source: AT&T.

and shopping services. Consumers can use many different access appliances: a television with a set-top box, a wireless telephone, a private branch exchange (PBX), and a local area network (LAN), for example. The various terminals are optimized for the talents and interests of the people who sit in front of them, but very few people can operate them at all.

AT&T's role is to make these instruments as friendly to use as the telephone. This objective is difficult. Even the task of putting telephony on a computer platform and making it quick and easy is not simple. Customers making telephone calls from computers must wait three minutes while their machines warm up and the modems start running.

Connecting the content players to customer premises are industries in broadband interactive telecommunications (two-way video, two-way voice, two-way graphics, and some computer intelligence). These industries include cable players, direct broadcast satellite players, competitive access providers, local exchange companies, and various wireless players, from cellular to those who plan to distribute cellular-based video from a wireless network. Those channels all provide ways to distribute content to telecommunications devices in homes and businesses.

Most cable TV companies are heavily in debt. They still owe money to the bank for the networks they have in place, and the networks need improvement because they frequently fail during storms. Most systems have only about 12 channels, and many companies want to increase the number of channels to 35 or 40. Now, they want to move into the broadband multimedia two-way interactive business, which requires a lot of money. To borrow this money, they must go to the same bank that already holds their debt.

While cable companies try to find new financing, DBS providers will provide increasing competition for them. Sometime in March 1994, General Motors, AT&T, and others will turn on the satellites they have launched. The satellites will offer between 25 and 125 channels each. Customers with satellite dishes will have access to a tremendous number of television channels, channels that do not go out with every storm as cable sometimes does. The satellite owners, which will have tremendous sunk costs, will try to gain market share by offering a low-priced package to compete with the cable companies. Today's cable users are the most likely buyers for DBS and will probably experiment with these new services. DBS companies may not be a great success, but their actions could be very damaging to cable TV.

Wireless has much promise and no real natural enemies other than new competitors trying to establish themselves. Still, it faces important challenges. Wireless companies need the capability to interact in a friendly way with other networks, allow secure transactions on the telephone, provide services at a reasonable cost, and allow billing to the calling party. On today's wireless networks, the cellular phone user must pay for both incoming and outgoing calls.

The CAPs are important as short-term players because they have fiber and can deliver broadband capability. They may not have a terrific long-term future, but those that are well run are attractive holdings today. Their strategy is to install fiber optic lines to communities, fill them up with a little traffic, and then sell them to a company interested in increasing its local access interests.

The LECs are also in an attractive position. They are not as helpless as they would have others believe, and they are making very bold moves. Their natural enemies are each other as they move into one another's territory.

Our goal is to stimulate a market in which each of the delivery conduits can mimic or reproduce the capabilities of the others. We want all telephone companies to have broadband, two-way interactive capability. We also want LECs and cable operators to have that capability and even want to see satellite groups have some kind of interactive capability.

Continued competition among these players will bring the costs of access down, which will help AT&T deliver services to its customers. Today, the local telephone companies collect all the money associated with delivering services to homes and businesses, but these other players want to enter that business. Most of AT&T's money will continue to go through the telephone companies for a long time, but we would like to reduce those costs, which are extremely profitable for the local telephone companies.

A world with multiple competitors providing access to the home would play to the complexities of life that AT&T works with. We do not want houses in the United States to have only one wire. That would leave us with a worse bottleneck than we have today. With or without regulators, homes will have a combination of wire and wireless technologies.

If we want all these delivery systems competing with each other with a chance to survive, we need to help them. We have offered to help cable companies, local telephone companies, and other players by working with them as their network integrator and builder. That way, we go with them when they apply for loans to build their networks and we can guarantee the networks will work when the loan payments begin.

AT&T is an established integrator. It has not yet, however, established this reputation with cable companies, even though it supplies that industry with much of its cable and electronics. In the near future, we believe they will ask us to rebuild their networks. We will have an opportunity to manage those projects, and we think we can make a lot of money as they build out their networks, which is strategically to our advantage.

AT&T has an opportunity to act as a host for content players. The content players' products will be distributed through all the different home receptacles to consumers with no real understanding of how the distribution networks work. We can offer the content players not only the capability of delivering their movies to the consumers' TV sets but also a way to deliver those movies through any of the other networks. We might even be able to operate data bases for them that can tell them who bought the last batch of movies. We would like to offer content providers a set of common services to store content in digital form and to bill for that content as it is distributed. Nobody has yet figured out how to meter content or how to keep the copyrights working. We will be able to provide security for that network and distribute the content over a wide area (so the producers do not have to "bicycle" tapes around the country).

Providing a whole array of common services makes sense. For example, for a computer software vendor, we might offer a vault in our network, a digital bank to store software. The vendor would have continuous access to that software without telling us, so it can be upgraded regularly. When customers want software, they dial just as they do today over any type of access device. We will optimize delivery of the software for whatever the devices are, send it to them, bill them, and handle the royalty reflow. The real value is that customers do not receive software that is instantly out of date. The economy we will offer to those content providers is eliminating enormous amounts of inventory in the chain of distribution to their customers. This is the world that we envision, although we do not know exactly how it will develop.

Conclusion

Today, people think they ought to get into the content or the distribution business, and they might be right. Many companies are spending a lot of money to buy content; the next thing they must do is pay off the debt. The question is whether they will have enough money to build a distribution system to deliver that content to the home.

For AT&T, now is the most exciting time in this country since the building of the railroad, because we are making real improvements in the infrastructure. The race is on, and AT&T is the best partner to help build that infrastructure. As a result, AT&T will make money as other companies spend money to build infrastructure. When the infrastructure is complete, AT&T will see if it can find another opportunity. The first round of acquisitions in this business will not determine who will make profits during the next ten years.

Question and Answer Session

Richard S. Bodman

Question: What role, if any, will the federal government play to help or hinder the building of this conceptual landscape?

Bodman: Currently, the government is playing catch-up. President Clinton and Vice President Gore have made it acceptable to discuss new networks such as the national information infrastructure. They have been helpful in breaking the logjam to start construction of the networks. In the very near future, AT&T and its major competitors will have in place everything needed for this national information infrastructure. It will, however, be built with private capital and no government assistance. The government has been useful in promoting the idea of security. We need to make transactions secure so companies can use the network and bill for content without fear of it being stolen as it travels across the network.

Regulators will focus on ensuring that content companies and long-distance companies have fair access to local communities at fair prices through competitive means. As we build these broadband networks into communities, we must be certain they are not controlled by a single entity. About 99 percent of traffic still goes through the RBOCs. That is neither good nor bad, but as long as that is the case, we must be careful not to let them bottleneck the broadband system.

Question: How would AT&T's international network strategy differ from its domestic strategy?

Bodman: I am not sure the international vision is exactly the same, but telecommunications improvements are occurring globally. I would watch people like Rupert Murdoch, who has major satellite programs in Asia, Europe, and England. They are still in the process of breaking down the international barriers, and satellites are the fastest way to do that. AT&T is positioning itself for the next several years to provide good basic personal and corporate services to our overseas customers. All of the excitement about broadband multimedia is evident, but it is secondary to expanding our basic core businesses worldwide.

Question: How does raising prices twice this year fit into AT&T's strategy?

Bodman: Each year, the overall price AT&T charges is lower, so we try to squeeze the last dollar out of some sectors while making concessions in others. Productivity gains in this industry are enormous and growing. A rising price structure for the long term is unrealistic. AT&T and its competitors are positioning themselves. We are trying to do better in certain segments in which we may not be covering our cost of capital, but fundamentally, we expect prices in this business to decline.

Question: Where is the greatest part of the value in the value chain?

Bodman: For AT&T, value has to do with talents, not just raw value. The real value is in building networks for other people and providing hosting services to enable software companies to address a common interface to get their programs to market quickly. They want to use these common services because they are not interested in billing and such. We think that is the highest value for us and for our customers.

Question: What is AT&T's strategy regarding the PCS auction?

Bodman: McCaw will lead our wireless capabilities and look at the PCS frequencies as we fill out our network. We have a big job ahead of us to make sure that the McCaw coverage is nationwide and seamlessly tied in with all of the other wireless players in the country. AT&T and all of its competitors will be present not just at the initial auctions but at the secondary events, when some players decide they do not want to build out their networks.

Question: Please comment on your investment in Unitel, the alternative long-distance carrier in Canada. Given its large losses, has it lived up to expectations?

Bodman: We are in the process of expanding our overseas capabilities. We would like to do this in partnership with the major postal, telephone, and telegraph carriers, at least in Europe and Asia, but every country is different. AT&T wants to be a major competitor against the conventional carriers in both Canada and England. Canada is a part of the domestic market for us. We will do whatever it takes to be a major player in that market, but it is impossible to build a telephone system from nothing and make money in the early years. Obviously, we will try to minimize these short-term losses.

Welcoming the Future: Bell Atlantic's Approach

Raymond W. Smith
CEO and Chairman of the Board
Bell Atlantic Corporation

> Consumers using telecommunications, television, and computing technologies today must tolerate inconveniences to gain access, control, and choice. Bell Atlantic hopes to eliminate some of those inconveniences and be at the forefront of providing easy-to-use services at reasonable prices.

The three basic modern communication appliances—the telephone, the television, and the personal computer—are merging into one, as are the providers of these services. This convergence is creating a major change for society and for what used to be separate industries—from the telephone to publishing. The proposed merger of Bell Atlantic Corporation with Tele-Communications (TCI), is a reflection of this convergence and change.

To some, convergence conjures up images of a Darwinian struggle over the same piece of ground, with the losers lumbering off to extinction. When Bell Atlantic looked at convergence, it saw the possibility of joining common visions. The attitude is best captured by the term "interactive multimedia." Indeed, the opportunities the change provides have forced some of the old forecasters of market share and future growth to lumber off to extinction because their forecasts were based on static assumptions and zero-sum games that no longer exist. For those companies and investors that have vision and have the determination to break out of old models, the world of interactive multimedia offers unprecedented growth and opportunities.

In this presentation, I will discuss three propositions:
- How new customer requirements have outgrown the boundaries of today's communications and entertainment businesses;
- Why Bell Atlantic believes the only effective response to new customer requirements is rapid, fundamental, industry-bending change; and
- How the transformation of Bell Atlantic from a legal franchise to a market franchise challenges investors to find new ways to assess and capture the potential of an essentially brand-new business.

Consumer Demands and the Need for Change

The two businesses that Bell Atlantic's proposed merger will combine—video entertainment and communication—are today served by companies that could not be more different. Cable is a profit-oriented business. The bulk of the cable industry is made up of small, entrepreneurial companies that operate with minimal bureaucracy and even less capital. That approach works fine if a company can grow by adding subscribers or buying more franchises, but it is not acceptable if competition is pressing the company to give better customer service, upgrade networks, and serve a more complex set of customer requirements than in the past.

Telephone companies, with their history of operating under rate-of-return regulation rather than competition, tend to be large, bureaucratic operations. They have lots of capital, and they spend it heavily—sometimes indiscriminately—in response to the perverse incentives of traditional regulation. That approach works fine if a company is the only game in town, but it does not work when competitors are targeting the company's prime markets and taking its best customers away.

As different as they look on the surface, today's cable and telephone businesses find themselves at the same decision point about their futures. In many ways, both industries have reached the limits of their original franchises and must look for new sources of growth and competitiveness. Both have mature technologies and customer bases, face competition from new entrants, and operate distribution systems that are limited in their ability to meet changing customer demands.

Not only do the two industries face similar risks but also they face tremendous opportunities *if* they can reinvent themselves to meet the enormous market demand for interactive, on-line, personalized, easy-to-use, easy-to-find information, communications, and entertainment. The demand for these services is huge. It is untapped and near at hand. Meeting that demand is the key to revitalizing cable and telephone services.

Consumers will go to amazing lengths today to get the choice, control, and convenience they require. They make their way through a confusing array of consumer electronics, odd combinations of devices, and primitive two-way capabilities. For example, to get on-demand programming, consumers climb into their two-ton vehicles and burn a few liters of fossil fuel to get to the video store. At the store, they pick out their fourth movie choice—because their first three are not available. Then, they go home and play it on their $300 VCRs, which they use *only* to play rental videos. (Some 80 percent of U.S. homes have VCRs, but programming them is a source of frustration in probably 100 percent of those homes.)

To access on-line services of other types, consumers must remember various personal identification numbers (PINs). They need one set of numbers to do their banking, another set of numbers to access E-Mail and Voice Mail, and so on. In addition to remembering their PINs, they must find ways to check their messages and maneuver through the various voice and computer mailboxes. For on-screen transactional service, they tolerate the crude interactivity of today's home shopping services. Consumers go through incredible inconvenience to buy (incredibly strange) goods by the billions; 22 minutes of a television shopping service can yield 15,000 sales. Customers' desires for shopping convenience have made services like QVC Network and the Home Shopping Network today's most elaborate Tupperware party. People are yearning to do things from home, so they put up with these inconveniences.

Similarly, to harness the power of real-time management information, the average executive juggles a fax machine; a voice mailbox; a cellular phone; a personal computer with a modem; and sometimes, a personal digital assistant, which may or may not recognize handwriting.

To satisfy a teenager's insatiable craving for Super Mario Brothers, parents may obtain a video game player with a 32-bit chip, a computer-disc ROM drive, and an add-on feature that uses digital compression to show full motion video—for only $600 a child.

Technologists are looking for a new application that will drive market acceptance of a new product, but the promising multimedia application is right under our noses. If 80 percent of Americans are willing to pay $300 for a machine they cannot even use properly simply to rent movies when they want them, they will surely pay for an easy-to-use, on-line service that gives them time-shifting and on-demand capabilities. People want to watch a New York Giants game when they want. The United States thus has a tremendous built-in market for an application that would offer interactive content, expanded choice, and user control, in an easy-to-use way and at a competitive price. The U.S. market for that application is about $15 billion and growing. The surprise, then, will not be a multimedia market that fails to materialize; the shocker will be how fast the market develops.

Neither of today's major distribution systems for this application—cable or telephone—is currently set up to deliver what customers want. Therefore, Bell Atlantic plans to deliver those services through its networks, software, and connections. The question is: How should Bell Atlantic go about exploiting these new growth opportunities to establish a competitive advantage in this new industry for the next decade and beyond? Deciding that the time for incrementalism is long past, that neither passive investments nor half-hearted demonstration products will suffice, Bell Atlantic has embarked on a total transformation. The proposed merger with TCI is part of that transformation; it is designed to make Bell Atlantic the market leader in what will be the largest industry to be created in our lifetime.

Reinventing Bell Atlantic

The combined strengths of Bell Atlantic and TCI would allow Bell Atlantic to do two basic things. First, such a combination would provide the ability to extend and strengthen existing distribution franchises. Specifically, Bell Atlantic could invest in full-service networks that enable telephone platforms to generate cable and video revenues and enable cable platforms to generate telephone revenues. This penetration of existing markets would have the potential to double Bell Atlantic's current revenue per subscriber. In turn, these high incremental returns could be used to speed up the development of tomorrow's markets—markets such as personal communication services and interactive multimedia.

The second benefit the merger would provide is expanded opportunities to participate in markets that will not only be high growth but will also transcend geographical boundaries. Global branding will be very important in the new and vibrant markets for video programming, intelligent network capabilities, proprietary intellectual property, digital file servers, operating systems, and so forth. Therefore, the ability to establish a true global franchise and brand identity is a primary purpose of the proposed merger.

Bell Atlantic decided that these benefits could not be achieved unless the entire industry reshaped itself, which is exactly what the proposed TCI merger is intended to do. We decided that providing products with the right combination of price, quality, and ease of use to create the new interactive multimedia market would be possible only by a company of the scale of the planned Bell Atlantic–TCI.

In addition, the merger would put Bell Atlantic in the position of being first in the market. A merger is the quickest path to early market entrance on a broad scale, and the earliest market entrant has a distinct advantage. For example, Home Box Office beat Cinemax into the marketplace by only a few months, but HBO has never lost its position as market leader, even as competitors multiply. Getting to the market first with a flexible network infrastructure and a desirable package of entertainment and communication services—wired and wireless—is the surest route to market leadership. Bell Atlantic's intention, which has also been behind its drive in video services since 1991, is to be the first to satisfy customer demand for interactive services, to gain market acceptance, and to establish brand leadership.

Several company initiatives would create value almost immediately. We will offer video to current telephone customers in some markets in 1994 and to the top 20 markets in our telephone territory by 1997. With TCI's cash flow and cable expertise, this regional build-out would be faster and more profitable than Bell Atlantic alone could ever accomplish. TCI would also immediately and dramatically increase its capital budget to build full-service networks in its territory. Bell Atlantic's plans following consummation of the proposed merger are to invest $15 billion through both companies during the next five years in profitable new investment opportunities in our combined territory, much of it mined out of the traditional telephone construction programs. These projects would all be funded by internal cash flows. This facet represents a quantum leap in capital efficiency, especially on the telephone side of the business.

Our plans are substantial. By offering cable to telephone subscribers and telephone services to cable subscribers, Bell Atlantic plans to generate the incremental revenues necessary to migrate quickly to new technologies, products, and customers. Cash flows from the new products and services are expected to begin having a significant impact on our business as early as 1995, with real momentum building in the second half of the decade.

We also plan to climb up the value chain toward offering a branded package of video and telephone services to deliver on the company's expanded network platform. An example is the initiative to create a single brand identity for all proprietary software products.

On-line interactive multimedia will not only create a whole new class of customer services but it also has the potential to make a radical change in the relationship between customer and supplier and to alter the structure of manufacturing, sales, service, and distribution.

To achieve its goals, Bell Atlantic must resolve a number of large issues. For example, the new products and services must be affordable, which will be achieved only when sizable companies order large quantities of product, spreading development and technology costs over a wide base. The scale advantages will be significant, but only those players willing to expand the size of their distribution systems dramatically will be able to capitalize on this advantage, not those who merely invest passively in someone else's existing technology or operations.

Multimedia products will also have to be easy to use. Bell Atlantic wants to give customers products with a comfortable look and feel that will become the world's standard, just as Windows is becoming the standard interface for personal computer applications. We are partnering with such companies as Oracle and Intel to develop server technologies, software interfaces, and data bases that are simple enough to let an 8-year-old order a pizza or an 88-year-old talk to a doctor and that are sophisticated enough to deliver entertainment and transactional services on demand to thousands of customers—together with the advertising, billing, and support to back up the services.

At Bell Atlantic's December 1993 analysts' conference, we will introduce a proprietary interface called Stargazer, which we believe could be one of the first and best video-on-demand navigational systems in the business. Bell Atlantic came up with Stargazer on its own, so we are excited about what we will be able to accomplish together with TCI, one of the premier programmers and packagers of video entertainment in the world. Thus, we believe the opportunity for us to invest in new businesses is tremendous.

A company that can reach millions of customers over an open network will stimulate the innovation and creativity that will truly usher in the multimedia age. We believe that the market for transactional and entertainment services will grow so large that no one company can hope to dominate it. In addition, customers have spoken out in favor of choice, and market diversity calls for vendor diversity. Therefore, the market will be best served by the presence of the greatest possible number of information service providers. Thus, Bell Atlantic wants to offer easy access to the new distribution system to multiple content providers and early adopters. We want to ensure the quickest possible ramp up of interactive multimedia services, and we want to provide entrepreneurs of the multimedia age that do not have much capital the opportunity to bring products to market.

To flourish in this marketplace, we must free the last vestiges of the monopoly mind-set. Bell Atlantic is fostering a world of competing networks, in which the only protected markets will be those created by excellence, early entry, and the strength of customer loyalty.

Valuing the New Franchise

The transition from a legal franchise to a market franchise not only challenges traditional wisdom about how to run a telephone company but also challenges traditional definitions of how to value one. In effect, Bell Atlantic wants to create a new industry, so investors will need to look at the company in a new way.

As we push into these new territories and seek to strengthen and extend our franchise, our accounting returns will be different from what they have been in the past. Therefore, the questions that need to be asked of the new Bell Atlantic are the ones asked of all growing, expanding franchises: Is the business expansion accompanied by an increase in revenues, in cash flows from the customers? And is the capital being deployed to generate high incremental returns that reward investors?

If the merger is completed, the proper measure of whether it is creating value for shareholders will be cash flow growth. The new Bell Atlantic should be valued, as other growth companies are, on a cash flow multiple. Looking at us through that lens, analysts will see a strong company. They will see cash flow growing at an annual rate of 10 percent or more. Almost immediately following the close of this merger, they will see a multiple of at least 8 times cash flow, which we expect to expand as we demonstrate our ability to sustain cash flow growth. They will also see a very strong financial position, as indicated by bond ratings of at least A.

Looking at us in that light puts the new Bell Atlantic in rare company. For comparison, we recently set out to examine U.S. companies with cash flow multiples of 8 and projected cash flow growth in the 10 percent range. The list was small but impressive: Merck & Company, 3M, PepsiCo, Schlumberger, Ltd., Wal-Mart Stores, and Home Depot—companies that fit just about anybody's definition of holders of a market franchise. We are confident that we have the strategies, the people, and the market presence to be one of them.

Since the AIMR Conference in November 1993, a number of shifts have occurred in the communications industry. Among the most dramatic is Bell Atlantic's decision in February 1994 to terminate merger negotiations with TCI and Liberty Media.

In the weeks since calling off the merger, Bell Atlantic has reaffirmed the strategic direction outlined in this presentation. Bell Atlantic remains committed to being a world leader in information, communications, and entertainment, and it will continue to pursue its three fundamental strategies: (1) building full service networks, (2) developing and delivering a robust package of entertainment and value-added information services over those platforms, and (3) expanding beyond its regional boundaries into high-growth domestic and international markets.

What has changed are the tactics Bell Atlantic will use to pursue its goals. The company still sees viable investment opportunities in full-service networks, as well as the programming and packaging of content, and it will seek strategic partnerships and alliances in selected high-growth markets. Given the dynamic state of technology, regulation, and industry structure, however, a single grand alliance on the order of the Bell Atlantic–TCI merger is less likely to succeed than a more flexible, market-by-market approach to business expansion.

— Raymond W. Smith

Question and Answer Session

Raymond W. Smith

Question: You have often said Bell Atlantic wanted to be the highway, not necessarily to own the content going over that highway. How has that position changed?

Smith: Bell Atlantic has always been in computer software; the need now for the industry is to develop interactive software. The intellectual software and intellectual property flow over the interactive software. We do not necessarily have to own the intellectual software. The programming assets we have are primarily programs and packages, not directed to the production of content. Periodically, Bell Atlantic may take a small equity position in a content provider in order to gain access for our network, but we see ourselves primarily as packagers, programmers, and facilitators throughout a broad area. In the next ten years, we will move up the value chain—not necessarily to be a majority owner of, for example, a studio but perhaps to take a position in it.

Question: How will video be delivered to the home?

Smith: Technically, Bell Atlantic can deliver video over the telephone lines in either of two ways. ADSL is a good interim technology that will probably still exist 25 years from now. It merely compresses a video signal. The 6-megabyte signal travels over a telephone line and provides extremely high quality. It requires a wiring board in the central office and one in the set-top. We are currently testing it in about 200 northern Virginia homes and are happy with the quality. The ability to manipulate digital signals is excellent even in the current rough and rudimentary ways of doing it. We will use that technology in areas where demand exists but is not yet at a high level of penetration. ADSL technology can be installed home by home; the whole neighborhood does not have to be wired. We can wire 5 houses out of 13 in a neighborhood. Then, when that neighborhood is ultimately wired with broadband, the second technology, we can reuse those circuits. We estimate those circuits will be used many times in their lifetimes, as other circuits are.

We are bringing broadband to locations in New Jersey and Alexandria, Virginia, in 1994. The 364-channel service will offer video on demand and do everything that the software allows it to do. At the same time, we will deliver telephone signals. Consumers can use the compressed copper or ADSL signal for telephone while simultaneously watching television over as many as three sets and have total control over those television sets in the home. We have various technologies and architectures to determine what is the best price and the best way to deliver video. Broadband is the way telephone data, video, and all terrestrial signals will be delivered in the next 25 years.

Question: Please explain how cable phone works.

Smith: Cable phone can be provided in several ways, one of which is to turn down a number of channels and provide an unsophisticated, U.K.-style, cable telephone service. Its operating system capabilities and intelligent network services would be limited, but it could be provided relatively simply in areas where penetration would be easy.

The alternative is to use a wireless tail. We would attach a 10-megahertz wireless bit of spectrum to the cable plant to bring fiber to what is called the node, which serves 200–500 homes. Then, we would use wireless into those homes in which wireless service priced at a premium for portability would be attractive.

Question: How does Bell Atlantic intend to defend its market franchise against competitors?

Smith: Bell Atlantic currently has one-seventh of the U.S. telephone service market. No doubt, we will lose market share with the changes underway in the industry. Competitors will emerge, and where we deserve to lose business, they will even use our switch to take service away from us as regulators require us to provide access to our facilities. Bell Atlantic is by far the lowest cost company in the world, however, and our cost curve continues to decline. We are probably not an easy target as a competitor, because our costs are low, our service is high, and we are reasonably responsive. So, although we will get nicked in certain places, we will continue to offer superior service at reasonable prices.

Question: Why should Bell Atlantic in the future sell at the same cash flow multiple as cable companies?

Smith: It is an evolutionary era, and ten years from now, Bell Atlantic will not look as it does today. The capital structure will change. Looking at the cash flow multiple of some of the cable companies versus the multiple that Bell Atlantic has today, analysts will realize that we have room to move up. That multiple can serve as an aspiration. We have the strategy to achieve it, but given our size, may not be able to sustain it for a long time. Nevertheless, our goal and plan is for the multiple to be up into the high teens.

Question: What issues for the next few years worry you the most?

Smith: Number one is development of the operating application software to bring the proposed services on line. Although we will be doing some of that development ourselves, some of it will be done by the software companies and will be out of our direct control. These companies are showing a great deal of interest in this project, however. Companies such as Oracle Systems are putting tremendous amounts of money into development. So, we hope they will be ready when we are ready with the networks.

The second concern is whether we can produce the proposed network at a reasonable cost. Will the cost curves of the various components of the network—the server technology, the fiber optics, and so forth—match what we have in our plans? People who are going to make a major investment must worry about whether they can count on a declining cost curve for capital expenditures.

I am confident of resolution in both these areas. I have been traveling abroad to the various software centers of the world, and I am impressed with the speed of issuance of generic software for the consumer interface. I am reassured that a user interface in operating software will be produced that will be reasonably priced and developed on time. As to component costs, some of the manufacturers are willing to agree to long-term contracts and low prices for some of the components of the network, which is reassuring.

Question: Who is on the cutting edge of interactive software?

Smith: I think the Media Lab at the Massachusetts Institute of Technology is on the cutting edge and has been for a couple of years. Nick Negroponte, the father of the Negroponte switch and some of the concepts that have become popular in the past year or so, illustrates that the interactive appetite of human beings is insatiable. He can demonstrate tremendous changes in behavior regarding technology use in very short periods if appropriate, user-friendly educational software is used. The Media Lab and similar activities at Carnegie Mellon and Stanford University signal that some of the early adopters will come from the university community and that we will not have to count solely on such companies as Microsoft.

Question: What are Bell Atlantic's plans in Mexico?

Smith: We are very excited about the Mexican market. We have a cellular opportunity in six of the Mexican states, and we also have nationwide fixed wireless spectrum. The government has given us permission to compete directly against Telmex not only in cellular but also in basic telephone service. We can provide this fixed wireless service in many of the areas where Telmex is not even deploying service.

Question: What legislative changes are you anticipating in the future?

Smith: I expect that positive legislation will soon come out of the U.S. House of Representatives. The legislation will include total and almost immediate elimination of all manufacturing restrictions, and in a short period, companies like Bell Atlantic will be permitted to engage in the resale of long-distance services, in cable, information services, and cellular. The facilities-based inter-LATA long-distance toll restrictions will be eliminated within a couple of years, and we will be free to enter the long-distance business on an unrestricted basis.

The Interactive Multimedia Market

Lawrence J. Haverty Jr., CFA
Vice President
State Street Research & Management Co., Inc.

> The interactive multimedia market is exhibiting the signs of a speculative environment. Investors can protect themselves using an analytic framework to look at both the business and the finances of telecom companies.

My view on this subject is that of a buy-side analyst. My firm has no banking relationships, and it does not prosper from trading securities. Our only business is to make money for our clients. We look for stocks that we can purchase at a reasonable price and that will increase in value. An analyst is very different from a reporter. A reporter listens and parrots. An analyst listens, thinks, and then draws conclusions.

This presentation will analyze some of the finer human frailties: fear, greed, avarice, and stupidity, all of which are visible in the interactive multimedia market in fairly large quantities. Raymond Smith said a lot is new, but from an analyst's view, nothing is new.[1]

History and Folklore

Twenty years in the business have led to the recognition of several signs of a highly speculative environment in this industry. We are in a dangerous environment for investors. The first sign is seminars. One in October was called "Interactive Television," and it cost $985.

The second sign of a speculative environment is crowds. For example, when Ben Rosen, now chairman of Compaq Computer Corporation, was an analyst at Morgan Stanley and Company, as few as ten people met for the first personal computer seminar. Now, such seminars, including this one on telecommunications, draw tremendous crowds. As a former colleague of mine said, "When you are sitting in the balcony, it may be time to head for the exits."

The third sign is parties. Parties signal trouble. Investors can buy and sell a stock based on the lavishness of company parties. Some of the famous peak parties include that given by Digital Equipment Corporation. When its stock was $192 a few years ago, it rented the Queen Elizabeth. The stock has since dropped to $29. Jerry Sanders at Advanced Micro Devices has thrown some fantastic parties through the years.

Another sign of a speculative environment is inventories. Some of these telecommunications companies actually produce products. The proclivity of a manufacturing business is to produce too much. These inventories are hidden on the balance sheets in the 10-Qs.

The next sign to watch is supply. Profitability is a function of supply and demand; new companies do not increase demand. A few years ago, I was convinced that a tremendous amount of money could be made when customer premise equipment in the phone business was freed from regulatory controls and users could buy their own phones. I went to the consumer electronics show and tried to figure out a way to make money from this idea. I asked someone where the items were being displayed, and I was told in the phone room. When I found the phone room, it turned out to be the size of two football fields. Consequently, I doubt anyone made money in customer premise equipment. The demand was growing, but the supply was growing faster.

Another sign of a speculative environment is jewelry. When analysts attend roadshows for IPOs, they should look at the watches the managers wear. Never invest in a company whose management owns more jewelry than you do.

The last sign is companies with an executive position of chief scientist. They want analysts to think they are doing something mysterious or that the analysts cannot understand. A chief scientist probably signals trouble.

---

[1] See Mr. Smith's presentation, pp. 73–76.

Analytic Framework

Analysts need to conduct their investigations in two phases: financial analysis and a business analysis. They search through documents to check up and down the supply chain. The only document worth having is the 10-Q, the quarterly SEC release. The 10-Q has a balance sheet, income statement, and cash flow statement. The game is won or lost in the cash flow statement.

The financial analysis has three facets: cash, inventories, and receivables. Cash is especially important when dealing with the film business. Most people would probably consider it fun to follow the film business, but it has some nuances to consider. Anyone can produce a movie; all that is needed is money. The average cost to produce a movie is about $20 million, with prints and advertising costing another $10 million.

Although the cash is spent immediately, accountants allow companies to defer the costs of production over the life of the product. No one really knows how many years the film will have revenues, because it has several distribution streams—domestic video, foreign video, domestic/foreign theatrical, and any one of several deals that could be done on television. As a result, companies in the film business can report earnings even while they are spending cash at an incredibly rapid pace. For example, the quarter that Carolco Pictures released *Terminator II*, which did a big business at the box office, the company was cash negative. In fact, it was cash negative for 14 quarters and was never cash positive. Carolco stock trades at less than 50 cents a share.

Analysts should also analyze inventory. One of the greatest inventory debacles was Atari Corporation. Warner Communications, which owned Atari, fully disclosed everything, but analysts had to look for it. On December 31, 1980, Atari's inventories were $73 million. By September 30, 1982, inventories had increased sixfold, which was far faster than Atari sales increased. At the time, Atari had a game called *ET* that never sold. From the peak, Warner Communications' stock lost about two-thirds of its value because Atari, a kind of interactive multimedium, simply produced too much inventory. They bought landfill to bury it.

Receivables can also be a problem. In 1983, Coleco, the developer of the Cabbage Patch Doll, was entering the video game business with Colecovision and Adam. It was making life miserable for Atari, the dominant firm in the industry at that point. Like Atari, Coleco overproduced its products, which were vastly inferior to other video games. What Coleco did that Atari did not, however, was ram the products down the retailers' throats so that the products did not appear on the Coleco balance sheet except as receivables.

When dealing with entertainment products, it is important to understand that the retailers have return powers. These return rights are called stock-balancing programs. The last thing entertainment producers want to do is field questions from analysts about stock-balancing programs. Analysts should watch the inventories and the receivables and look for year-on-year changes. Most companies will cleverly provide the end-of-year balance sheet and then the September balance sheet, but analysts need year-on-year figures.

Although the 10-Q is an analyst's best friend in a speculative environment, his second best friend is tracking end-product movement. A company in the interactive multimedia business recently showed the usual statistics about how fast the market is growing. This year, it said, the market is growing at a 50 percent rate. The stock continues to rise, so the market must believe it, but I cannot find retailers selling video games at anywhere near a 50 percent growth rate. Some signs of manufacturing backup are developing because Nintendo and Sega Games had some disappointing top-line results.

Characteristics of a Good Equity

Warren Buffett set the framework for analyzing a business for investment purposes: in equities, investors are business owners. Certain conditions characterize a good business or equity. One is that the business is growing. Two, the business should generate cash sometime during the investor's lifetime. Three, the business should have barriers to entry; if an investor wants to buy a lemonade stand business and ten lemonade stands are located across the street, it is probably not a good business. Four, the business should have little or no government regulation. If Washington is in the business, it denigrates the quality of that business.

People in the cable business tell investors that sometime in 1994 their operating cash flow will probably be down on a year-on-year basis; at the very best, its growth is decelerating. Currently, the momentum players are active investors in the cable industry, and they will eventually discover the cable business is slowing fairly significantly.

The stock market is currently ignoring some of the things that are occurring in Washington, such as a bill that is supposed to regulate cable prices. According to how government keeps score, the mean price increase in cable television has been about 7 percent a year for the past ten years. The CPI for medical care is also about 7 or 8 percent a year. The overflow crowds have disappeared from medical care conferences these days because the government has become involved and is trying to reduce the rate of inflation in medical care. Even with the government's price control programs, the cable TV CPI increased 8.1 percent in September and the medical care CPI increased 5.8 percent. The government is getting more involved in cable, which is not good for cable stocks because the government wants lower cable prices.

The Multimedia Business

Interactive multimedia is nothing analysts have not seen before. A good analytic framework, as seen in **Exhibit 1**, simplifies the various factors and allows the analyst to get around the hype of the business. Sometimes a manufacturer deals with the wholesaler, who ships directly to the retailer, and sometimes the manufacturer goes directly to the retailer. In the end, the consumer ends up buying the good or service.

Exhibit 1. Interactive Multimedia Analytic Framework

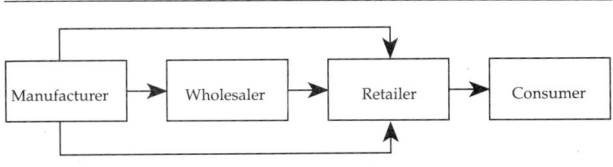

Source: State Street Research.

The two distribution channels for goods are trucks and rails. These carriers are commodity businesses. They do not generate superior rates of return for the owners except with very effective management. Conrail is an example of a rail business that has been very well managed. Inherently, however, these businesses do not generate superior returns.

Manufacturers in multimedia include Warner Brothers (films), Electronic Arts (video games), and PolyGram (records). Wholesalers are packagers of products manufactured by other people such as MTV (music), Showtime, and HBO and Company (filmed entertainment). Retailers include Blockbuster Entertainment Corporation, Warner Brothers Studio Stores, and the Disney Stores. The transportation networks link these channels.

Multimedia is essentially a service transportation mechanism with two competing entities: cable and telco. In five or ten years, most homes will have two competing services: one by the telephone company and one by a cable company that may or may not be owned by another phone company. The probability is high that returns will not be superior in that business.

Demand and Supply

Investors want to determine which multimedia businesses will have superior returns. Using an analytic framework, they can analyze demand and supply in the various markets. Economics students are taught the scenario of a constrained optimization model, in which demand is constrained by some variables. Two simple constraints on demand are time and money. In conversations about demand, however, analysts rarely hear constraints mentioned.

Analysts should consider the demand for multimedia. On September 13, 1993, *Advertising Week* reported that 55 percent of the public is ambivalent about interactive multimedia. In an exclusive survey on October 11, 1993, *Advertising Age* reported that only about 19 percent of the public is even aware of the concept of interactive multimedia.

Whether the demand will be fully effective is another consideration. Assume that a household pays $20 for basic cable service and $12 for a premium channel, orders maybe ten pay-per-view films a month, watches the Metropolitan Opera once a month for $12, buys a $10 or $12 game channel, and pays $25 a month for pay-per-view sports events. That adds up to about $120 a month.

My children are teenagers, and the 15-year-old enjoys cable TV. Heaven forbid if he had unlimited access to the television of tomorrow. The hypothetical $120 a month bill would probably be an opening bid. The bad news for these companies is that he would probably only do it for one month, because then the service would be pulled out.

Much of the demand for interactive multimedia will develop during my children's investment career, which is two or three times longer than our clients' investment horizons. Strong growth for interactive multimedia is expected during the next 20 years, but investors are interested in the period from 1994 to 1997. In the pension management business, for example, the most important factor is three-year performance, because that is how managers are compensated. Most of the growth, however, will be well beyond 1997.

Supply can be an important factor in this business. A good equity has effective barriers to entry. In interactive multimedia, Toys 'R Us currently supports about 40 hardware systems and more than 2,000 video games. The competition is tremendous. Even with effective entry barriers, the capital cycles can be fairly devastating to investors when the cycle turns.

The film business has probably as effective entry barriers as any industry; even so, the supply is increasing. One new major company—Savoy Pictures Entertainment—has been financed. It has raised about $500 million, and it can borrow to make films. The Walt Disney Company is doubling production, from 30 to 60 films a year. PolyGram, which is funded by Philips Electronics and has an unlimited supply of money, is also entering the film business. Orion Pictures Corporation is back with *Robo Cop III*. MGM plans to refinance. The film business is a very good business and tremendous capital flows are taking place. Only so many people can make good movies, however, and everyone wants to hire them. Their salaries go up, and film margins go down. Investors beware!

Capital Spending

In addition to demand and supply, analysts should watch the timing of capital spending. In the paper industry in the 1970s, one company after another announced its capital program. The stocks rose for a couple hours, and then investors would think maybe this capital program would not have a high return. As a very successful investor once said, "Sometimes it is better to invest after the capital has been spent." An industry may spend a massive amount of money on capital, but the return might not be adequate. It may be best to stay on the sidelines. Too much capital is entering film production. Investors beware!

Security Selection

Our firm has a number of principles we apply in selecting securities. First, be totally blind as to whether the security is foreign or domestic. We will buy a foreign security if we can get faster growth at a lower valuation. We have made significant money in Grupo Televisa and News Corp. These are Mexican and Australian companies, respectively.

Second, have a useful mechanism for taking into account the vastly different financial structures in this business. For example, Disney is totally unleveraged, but Comcast Corporation is leveraged in every way imaginable.

Third, have some mechanism for valuing hidden assets, because they are important in these companies. News Corp. has increased from $12 to $55 in the past few years because of its ownership of British Sky Broadcasting (among other companies), which is not consolidated.

Fourth, look for the sources of equity value. If it looks, quacks, and walks like a duck, it probably is a duck; however, if it only looks like a duck but does not walk and talk like a duck, then it is not a duck, and analysts should find another way to value it. For example, one of the most successful equities in this industry has been Comcast. Comcast has increased 20 times in value in the past ten years. On an apples-to-apples basis, however, Comcast's operating cash flow grew more than 10 percent in only four or five quarters. The stock appreciated faster because of the aggressive use of financial leverage in an environment in which interest rates declined secularly. Comcast bet the ranch and won. Sumner Redstone bet the ranch and has made billions of dollars in Viacom. Viacom was not covering its interest on a cash flow basis when Redstone took control of the company.

Bell Atlantic Corporation wants to be a duck, but it only looks like duck; it is not walking or talking like a duck. Bell Atlantic is doing several things that the companies that have been good equity investments are not doing—for one, paying taxes. Comcast abhors taxes; John Malone of TCI abhors taxes. Also, these companies pay only minuscule dividends. If analysts try to value it as a duck, someone will pick their pockets.

How an Institution Invests

State Street Research is a plain vanilla organization. It has $22 billion under management, half in equity and half in bonds. It is driven by fundamentals and has a growth bias. We like to regard ourselves as a long-term, multiyear investor. In the main part of our business, when one account buys a stock, all do. We place a very high premium on consensus thinking. Last year, we committed a lot of resources to figuring out what is going on in this industry. At one point, we had five people devoting the majority of their time to this task. We talked to several companies—Bell Atlantic, U S West, and AT&T—that were very helpful to us.

From these discussions, we developed our beliefs and then put a lot of money to work in a fairly decisive manner. One, we thought there would be more than one winner; we spread out the money. Two, we wanted to be involved with companies with manager investors. We loved TCI because of Malone. We loved Comcast. We were big investors with Viacom. We have always been big believers in Steve Ross and Bill Gates. Third, we believed that the owners of content would be the winners. Eventually, everybody recognized they were going to be the winners. Too much capital came into the industry, and we have taken profits. Fourth, the only important factor in this industry from a valuation viewpoint is cash flow. Fifth, we needed to consider our valuation mechanism for hidden assets. Sixth, when the market was shooting down brands, we thought brands were very important. For example, we have made a lot of money with Blockbuster. Blockbuster is one of the best brands in the United States. It is mindboggling to realize Blockbuster did not exist seven years ago. Looking at consumer preference polls, consumers prefer Blockbuster to their cable companies. Anyone who thinks cable will put this company out of business soon has another think coming.

Occasionally, we use unconventional approaches. *Jurassic Park* was a great experience in a crowded theater with the ultimate in sound. It would not be such a great experience on household television. The issue is what Jeffrey Katzenberg of Disney calls the communal experience: people must get together at one point in time to maximize the value of the product. *Jurassic Park* was very successful and paid for itself in six days. In the ultimate communal experience, Notre Dame will play Florida State for the football championship, and tickets are selling for $3,000. We are looking for places where the benefits of this interactive multimedia technology are being shared in communal experiences.

This approach is somewhat unconventional, but we think that people who can successfully market communal experiences will be winners. We have, for example, a lot of money invested in gaming. I would urge anyone to go to Las Vegas and see what Steve Wynn is doing to the Mirage and Treasure Island, look at what Caesar's World is doing, and look at what Circus Circus is doing. These companies are using the technology of multimedia and audience interaction, and they have built very good businesses.

We occasionally sell stocks. Our current holdings in this industry are roughly half of what they were. Sometimes when we sell stocks, we buy them right back after a couple of weeks. We did that three times with News Corp. It is not planned, but we are not the smartest people in the world. We are, however, smart enough to know when we are wrong, and we will reverse ourselves.

Valuation is the most important element in the discipline of selling. Three years ago, when Time Warner was heavily leveraged, investors could not give it away. Time Warner's cash flow growth has not been what people thought it would be three years ago. It sold minority interests in its entertainment businesses. Somehow, analysts must account for those interests. Time Warner reached the outside parameters of our valuation targets, so we sold it.

Fundamentals are also important. Five years ago, the chief financial officer of Disney said the most important development for the history of the company in the 1990s is EuroDisney. At its opening, the stock was 140 French francs, and the parties on the Seine were unbelievable. Although Disney had been our largest holding, we had a real question about whether EuroDisney was viable financially, so we sold the stock. The stock is now 38 French francs. We like Disney and Time Warner longer term, but the valuations were not there in the one case (Time Warner) and the fundamentals changed in the other (Disney).

Conclusion

These are interesting but dangerous times for investors in these industries. Signs of speculation are everywhere. By exercising discipline in analysis, investors can distance themselves from the crowd.

Question and Answer Session

Lawrence J. Haverty Jr., CFA

Question: How do you think Bell Atlantic expects to convince the market that these promised multimedia revenues will be realized? What is your opinion of its position?

Haverty: I think the proposed transaction with TCI is a good deal for Bell Atlantic because the cash flow multiple is not excessive and it includes a critical mass of cable subscribers. TCI has twice as many subscribers as anyone else, and that is very hard to replicate. Also, John Malone will be a tremendous asset for Bell Atlantic. At some point, this stock will be a great performer; I believe the future will happen. I do not think it will happen in the next couple of years because of the dilution of earnings and huge capital spending. The investment impact of the capital spending is uncertain, but Mr. Smith has to run this company for the long term. Shareholders are a powerful constituency, but they are only one constituency. I give him great marks for his vision, and sometime between 1995 and 2000, Bell Atlantic will be a great stock. When people see the short-term impacts, however, they may not want to jump right in. We are not investing in Bell Atlantic and U S West for these reasons.

Question: What is the future of wholesalers as you have identified them in the context of this expected proliferation of multimedia products?

Haverty: Not very good. Once consumers have near video on demand, they will quickly discover that Showtime and HBO have marginal use. If they are to be businesses, and they are both moving in these directions, they must try to position themselves essentially as incremental suppliers of products. HBO is trying to increase the supply of films and has been a major investor in Savoy Pictures. It is also producing some films of its own. I am not so sure that HBO and Showtime are great businesses. I would put a very high cash flow multiple on them.

Question: How successful do you think a company like Sega will be in distributing its games over cable?

Haverty: Distribution is the big question in video and films. Sega has built a very successful business selling through Toys 'R Us and other stores. Games are about 18 percent of Toys 'R Us revenue. Sega must be very careful not to offend the powers at Toys 'R Us, because it may lose the money it is generating from the retail channel. I am not convinced that this cable distribution is a terrific idea for Sega; it must be careful.

The film companies understand this issue. Where they get their revenues is critical. A typical film costs $30 million, and maybe it will do $25 million at the box office, of which $12–15 million is returned to the producer. A typical film will also sell to the domestic cassette market in the neighborhood of 250,000–300,000 films at $60 a pop, which is about $20 million. That $20 million the studio gets from Blockbuster and its friends is money in the bank right now. Studios are not about to disenfranchise that distribution channel. The United States has 60 million cable television households. The biggest penetration in any pay-per-view product has been 5 percent. To replicate that video channel at $4 per viewing, half to the film company, half to the cable operators, to get the $20 million, they must get 10 million households to watch the film, but that is a 15 percent penetration of television households, and no product has ever done even 5 percent.

Companies are making films at the rate of about one a day. About 250 films will be produced this year. Managers must watch where the revenues are generated. As they build up the next distribution channel (the television of tomorrow), they must be very careful that they do not ruin current distribution channel life. It is tricky, and these migrations will occur at a glacial pace.

The U.S. Cellular Communications Industry

Dennis H. Leibowitz
Senior Vice President
Donaldson, Lufkin & Jenrette

> The U.S. cellular communications industry has enjoyed tremendous growth as a result of increased competition and improved technology. A series of mergers, combined with new cellular services, will fuel additional competition and growth.

Growth of U.S. cellular services as measured by penetration has been phenomenal; yet the industry is still in its infancy. Forecasts of future penetration are constantly being revised upward. The performance of the stocks in the industry, however, has a checkered history.

AT&T developed the initial mobile communication system with one-tower broadcasting. The system had limited capacity because each channel could carry only one conversation throughout the city. After several years of testing, the FCC allowed a cellularization process that divides a city into several different cells. The capacity of the transmission within each cell is limited; thus, the cellular system reuses the frequency and increases capacity.

Although most countries award cellular licenses on a national basis, the situation in the United States is unique: licenses are available only for local market areas. In 1982, the FCC initiated a licensing process to permit cellular communications in urban markets. Each market had two licenses: wireline and nonwireline. The FCC gave free licenses to local telephone (wireline) companies for their territories. In some areas, two or more independent telephone companies shared licenses, but they received their licenses without cost. The second competitor in each market was a nonwireline operator. The FCC encouraged operators to compete for the nonwireline licenses, and initially the commission held comparative hearings for those licenses in the largest markets. The basis for awards was often marginal; winners sometimes received the licenses for covering as little as 50 feet of additional territory.

Because of the difficulties and time involved in comparative hearings, the FCC decided to hold lotteries for nonwireline licenses in the smaller markets. Many lottery participants joined "settlement groups" to combine interests with other participants in order to improve their chances of winning at least a partial interest in a valuable license. The process resulted in a fractionalized ownership structure.

Early in the licensing process, the FCC had to determine whether to allow wireline telephone companies to buy nonwireline interests in markets where they did not provide regular telephone service. For example, Bell Atlantic Corporation could not hold both cellular licenses in Pittsburgh, but whether Bell Atlantic could hold the nonwireline license in Chicago was unclear. As a result of a favorable FCC ruling, the telephone companies began to acquire nonwireline licenses in other markets. For example, U S West acquired a nonwireline license in San Diego; PacTel Corporation (now Air Touch Communications), the wireless business of Pacific Telesis Group, acquired one in Atlanta; and Southwestern Bell Corporation acquired nonwireline licenses in several markets, including Chicago, Boston, and Washington, D.C.

Because of their strong capital bases, this competitive trend changed the demand and financing sides of the equation and is important in analyzing the prices of systems. Over the years, telephone companies have bought out about half the nonwireline licenses. Because they also have not sold any of the wireline half they own, they cover approximately three-quarters of the total industry's coverage. The situation, in which a small number of telephone companies are competing against each other in different markets, is similar to what is evolving in the cable industry.

The U.S. cellular industry has tried to overcome fractionalized ownership and create a national license through a series of mergers. The closest form to a national network in conventional cellular would be McCaw Cellular Communications, which serves about 25 percent of the country's population.

McCaw has tried to make up for the lack of a national network and would be helped by its proposed merger with AT&T. Under the terms of its North American cellular network, McCaw–AT&T could offer seamless roaming service with automatic call forwarding. A customer dialing an associate's number in New York would automatically be connected to the correct party even if that party is traveling in Los Angeles. The merger activity, combined with the strength of the business, has fueled an escalation in prices of cellular systems.

The Growth of Cellular

At first, it was not clear that cellular would be the wonderful business it has become, and few people believed it would turn out as well as it has. The first systems under the U.S. award structure generally began construction and initial operations in 1985. (Two experimental systems—one developed by AT&T in Chicago and one developed by Motorola in Washington, D.C.—existed before the award structure.) Only two operators bid for licenses in the cities of Chicago and Boston. In contrast, in later years, sometimes thousands of people applied for licenses—even for rural areas in states like Montana.

Penetration forecasts have increased steadily. In 1985, AT&T assumed that cellular penetration after five years would be about 1 percent of the population. In 1990, however, penetration was 2.5 percent, or 2.5 times what anyone had expected. By the end of 1993, subscribers will total about 15 million, or 6 percent penetration. In comparison, penetration for conventional telephones on a per person (rather than per household) basis, including all business and residential lines, is about 60 percent.

The relative youth of the industry understates the success of the business. Many systems in smaller areas were awarded their licenses by lotteries in the past few years and have only recently finished construction and begun operation. Those areas represent a small part of the country.

Neither the increasing size of the U.S. industry base nor the economic recessions of 1990 and 1991 have slowed industry growth during the past few years. The Cellular Telephone Industry Association, an industry trade organization that keeps statistics on the business, reports that the number of subscribers increased 51 percent in 1990—the depth of the recession. The number of subscribers increased by 43 percent in 1991, and by 46 percent in 1992, perhaps because the economy began to improve. In the first nine months of 1993, the number of subscribers grew 40 percent.

Despite this substantial growth, the industry is still in the early developmental stages. Thus, companies now expect about 15 percent penetration by the end of the decade, only 15 years after the introduction of cellular. The actual penetration could be higher because of increased competition.

The United States has the second highest penetration in the world, behind Scandinavia, where cellular service began in 1982. Penetration throughout Scandinavia is about 8 percent (and higher in Stockholm and Oslo). Most other countries have lower penetration rates because the industry has generally been less competitive abroad than it has been in the United States. Given the elasticity of demand for cellular phones, the penetration rate in Scandinavia is an auspicious sign of what might happen in the United States.

The cellular industry has had substantially better growth in the United States than in other countries, particularly the European countries, even though countries such as the United Kingdom and Scandinavia have had cellular service as long as or longer than the United States. One reason for the slower growth in the United Kingdom may be that its recession was worse than the one experienced in the United States, but the major reason for the difference in growth rates is the competitive situation in the United States.

Operating Characteristics

Even though the industry is a duopoly in U.S. markets, the competitive situation has caused prices to decline. Various local markets have different competitors, with varied attitudes, and companies appear to be willing to take chances with prices to see what will happen. U.S. customer prices have dropped 30 percent in nominal terms. The average monthly bill of $100 has declined to less than $70 during the past five years.

Although prices have declined sharply, margins have held or risen, and profits have increased sharply. Volume and the tremendous incremental flow-through of additional business are responsible.

The cellular business has healthy margins and is very attractive economically. Once the hardware and the system have been established, incremental operating expenses are minimal. The flow-through of additional revenues to cash flow before marketing expenses can be as high as 80 percent. Some companies have gross profit margins (before marketing costs, depreciation, interest, and taxes) in the 70 percent range. Generally, such companies have achieved those levels of flow-through because of economies of scale in amortizing overhead, billing costs, and so forth. In highly developed companies, overall margins after marketing costs are in the 40–45 percent range and growing.

Controlling marketing costs is the key to managing cost structures to meet the competition's prices, and cellular companies have successfully reduced them. Historically, a start-up cellular company, lacking a large sales force, paid commissions to independent agents to market the cellular service throughout a city. As the industry evolved, the use of commissions, which were as much as $400 per activation, decreased. Companies built their own sales forces to handle most of the business and paid the salespeople a salary plus bonus that was less than the commission rates.

Cellular service plans have increased margins and profitability by tailoring sales commissions and marketing costs to usage patterns. Most cellular operators offer bundled minute plans. Under these, the customer pays a monthly fee ranging from $29.95 to $99.95 that includes a certain number of "free minutes," and the company charges for additional minutes at a per minute rate. Companies vary the per minute rate according to the monthly fee; that is, higher monthly charges have lower per minute rates and lower monthly charges have higher per minute rates. The bundled plans have helped reduce marketing costs per subscriber, which is another reason why cellular companies have been able to show growth in profitability and margins despite a 30 percent decline in prices during the past five years.

Capital investment per subscriber is also declining. Investment per subscriber is highest at the beginning of operations because companies must invest in the infrastructure to cover the entire market before starting service. After that, they can add radio channels to cell sites as demand warrants. If the coverage is sufficient and they can get more capacity out of digitalization and other technology improvements, they may not have to add more cells.

Cellular Technology

The approaching conversion from analog to digital systems will substantially lower the cellular industry's investment per subscriber. Hardware companies are now launching competing digital systems, even the most conservative of which provides a 3-fold increase in capacity per channel; the most ambitious system claims as much as a 20-fold increase. From a capital standpoint, this technology will enable companies to earn a good return on investment even at lower prices.

Four different technologies are available to the industry: the existing analog system and three competing digital standards. One digital standard, time division multiple access (TDMA), was first introduced by McCaw Cellular and Rogers Cantel Mobile Communications in Canada. It divides an existing channel into three time slots. The second digital approach is code division multiple access (CDMA), a packet switching concept that uses intervals of unused spectrum, such as pauses in a conversation. The packets are put together, sent out, and reassembled at the other end. PacTel, U S West, Bell Atlantic, and others claim a 20-fold increase in capacity using CDMA, but from the standpoint of technological development, CDMA trails behind TDMA. The third standard, FDMA, is a frequency hopping technique.

Some industry experts are concerned that the technology war in cellular resembles the beta-versus-VHS battle in videocassette recorders and that the winner may not be the better system but the one that got there first—in this case, TDMA. For long-term growth, the industry must resolve the issue of the competing digital standards.

Cellular Services

Two new kinds of cellular service are emerging to compete with the established wireline and nonwireline operators: ESMR and PCS. In addition, cellular companies are expanding into paging and data transmission services.

Enhanced Specialized Mobile Radio

As the name indicates, ESMR is an enhancement of specialized mobile radio (SMR). Historically, the FCC allocated a series of frequencies near the cellular band to SMR operators that provide dispatch to taxis, fleets of trucks, or point-to-multipoint services. The structure resembled the original mobile service (IMTS) with one large tower broadcasting everywhere in town. The capacity is limited by the number of frequencies, because there can only be one user on a frequency at a time. The frequency suits dispatch services, which require less capacity because most conversations are short bursts of information.

Historically, the FCC fractionalized the dispatch spectrum among dozens of operators in a market, but in 1991, Nextel Communications (formerly, FleetCall Communications) asked the FCC for permission to cellularize dispatch frequencies and develop another cellular system. Nextel successfully acquired as much spectrum as possible, particularly in the largest markets, before announcing its plans. The company established a block of spectrum (maybe two-thirds the size of cellular in some of its bigger markets) and received FCC permission to build the system. Its initial markets will be Los Angeles, where the system is currently being tested, Chicago, Dallas, Houston, New York, and San Francisco.

Nextel is currently testing among dispatch customers; the system is only point to multipoint and does not yet interconnect with the telephone system. In January 1994, Nextel began competing with the cellular industry in Los Angeles.

Nextel uses a TDMA technology. Its digital system offers sixfold capacity increases over analog cellular, which somewhat offset the more limited capacity Nextel will have compared with the cellular operators. Initially, Nextel will offer services at a price similar to that of regular cellular service. To differentiate itself, the company will offer customers multiple services—paging, dispatch, and cellular. In Los Angeles, which is the most successful cellular market in the country, Nextel is trying to capitalize on the fact that the Los Angeles cellular system has periodic capacity bottlenecks at major frequency intersections. Nextel hopes that offering digital quality with guaranteed access and no blockage will lure people to switch carriers.

Nextel successfully acquired capital and backing from Motorola (which is manufacturing the base stations and handsets) and outside investors such as Comcast Corporation (one of the large cable companies). With this financing and the success it has had in the stock market, the company has acquired several other SMR companies and is trying to build the infrastructure of a national network.

The ESMR providers are trying to create an advantage to offset the fact that they will initially not offer roaming on highways between cities. The ESMR industry has quickly folded into three companies. Nextel, through acquisitions and development, now covers two-thirds of the country. Dial Page operates in the Southeast, and CenCall Communications in the Rocky Mountain and northwestern states. Together, these three companies will provide a seamless ESMR network. In effect, they are building a third, fully competitive cellular system.

Licensed third providers are quite common in Europe and the rest of the world, but until the ESMR companies become operational, the United States has only two cellular competitors. In most foreign markets, depending on the nature of service and the spectrum, the third licensee can, in time, capture as much as a third of the business.

Nextel's business plan foresees the company obtaining 20 percent of the growth in the market. That percentage would result in only about 2.5 percent overall penetration of the country by the end of the decade, or about 10 percent of the enhanced cellular market.

Nextel is still in a speculative stage; it must spend billions of dollars on infrastructure. The stock market has bought Nextel's story, however, and has shown a lot of hope for the concept. Nextel's stock has performed better than any of the cellular stocks. The company has a market capitalization of just under $6 billion, with $40 million in sales from its traditional dispatch business. Some consider Nextel's per pop (a unit of population) valuation a measure of its attractiveness, but more than two carriers are competing in the market now, and investors must consider the construction costs.

Personal Communication Services

The next group of competitors are companies providing personal communication services, sometimes called personal communication networks (PCN). PCS is a miniaturized cellular system. It is not a new technology; it is a downsized version of cellular. The smaller cells and higher frequencies of PCS reduce the distance each signal is carried and thus require more cells for coverage. Small, light handsets and long battery life require the system to have low power, which also requires more cells. Although the PCS cells may be much cheaper than standard cellular cells, the overall cost for a system could match or exceed that of a cellular system because many more cells are needed.

PCS began in the United Kingdom in the late 1980s. The British government decided the two existing U.K. cellular carriers—Cellnet, controlled by British Telecommunications PLC, and Vodafone Group PLC, controlled by a separate public company—were too profitable and needed competition. At that time, cellular phones were used mostly by business executives in their cars. The government decided to license a new service of portable phones for nonbusiness consumers and established a licensing process for personal communication networks. The announcement of the new service in 1989 marked the top of the cellular stock market in the United States. Investors recognized that if competition was going to occur in the United Kingdom, it would probably also occur in the United States. That judgment turned out to be true, and it has taken the cellular market about four years to recover from the weakness that followed the 1989 announcement.

The British government allocated 1.8 to 2.0 gigahertz (GHz) in the spectrum to PCS and awarded three PCS licenses. (The cellular business in the United Kingdom operates at 900 megahertz; in the United States, the existing cellular business operates at 800 MHz.) One license has been returned because of the questionable economics of the venture. Another license, owned by Hutchinson Telecom, is reportedly for sale, but the facilities are under construction and scheduled to open in the spring of 1994.

Mercury One to One, the only system that is operational, began in September 1993 on a much reduced basis from what had been promised to the government four years earlier. The system covers only London, although it will expand as economics warrant. It is a joint venture of U S West and Cable and Wireless PLC.

Mercury began operations with pricing 30–40 percent lower than the existing cellular service. The cellular carriers, because they cover not only London but also the whole country, have the advantage of pricing flexibility over PCS competitors. To preserve their profits on their most profitable customers, the cellular companies responded to the lower prices, with special packages for London area consumer users. Revenues per subscriber per month in Britain declined, but the price base in the United Kingdom had been one-third higher than in the United States anyway, so this price decline brought U.K. industry revenues down from about US$100 a month closer to a level closer to that in the United States.

The competitive situation for PCS is more complicated in the United States than in the United Kingdom. Under FCC rules, existing U.S. operators have a flexible spectrum allocation. They can offer whatever services they like on their existing frequencies. They can match competition from new services as long as their digital capacity allows it. If they have the capacity, they can offer PCS at whatever price they want.

Several problems delayed the introduction of PCS in the United States. First, the FCC had no available spectrum. It attempted to solve the problem by allocating frequencies similar to those being used in the United Kingdom and Europe. About 50 percent of those frequencies, however, are occupied by microwave users for intercompany communications or by public-safety agencies such as fire and police departments. Prospective PCS operators must pay off those that currently occupy the frequencies. Private microwave users will be forced to move to other frequencies during the coming years. Public-safety agencies are not required to move, but if they need money, they will. This problem adds to the costs PCS operators will incur.

The bigger problem is that the U.S. government has decided to award the licenses by auction rather than by comparative hearing or lottery. Winners of the auction will pay an up-front fee for the license. Depending on the competitive environment, this fee could be a significant expenditure, and it must be considered even before building costs.

Compounding the other problems is the government announcement, in its market allocation rules, that each market will have up to seven new licenses of different sizes. Two licenses of 30 MHz bandwidth will cover areas larger than the existing cellular markets. In addition to these two licenses, each market will have one 20-MHz license and four 10-MHz licenses for the local trading area. The rules state that one of the smaller licenses will be allocated to minorities or small businesses and the two existing cellular operators will be permitted to bid on the others.

The experience in the United Kingdom proved that more than two PCS licensees are not economically viable, so an amalgamation of the U.S. licenses will occur. The FCC will allow any of the new PCS operators to accumulate up to 40 MHz of spectrum. (To put this in perspective, existing cellular operators have 25 MHz and can buy another 10 MHz.)

All of these factors will drive up the costs and prices of personal communication services in the United States. The government estimates that the cost of bidding will be about $10 billion, or $40 per pop. Building the system will cost another $20 per pop, and investors will have to add in the cost of buying out the microwave users. The total is a fairly expensive up-front investment for a service that will have substantial competition and will, therefore, require low customer rates.

If the existing cellular industry had to have competition, the companies could not have asked for a better situation. The stock market reaction to these government policies was mixed. When the rules were announced, cellular stocks remained neutral. Stocks of SMR companies, however, with their higher costs for building systems, increased sharply. The marketplace perceived that the government rules would delay PCS and make PCS operations less profitable than they could have been. Given the government's auction process, originally scheduled for May 1994 but unlikely to meet that deadline, and the necessary construction timetable, PCS systems will not begin operation before 1996. This delay allows SMR and cellular systems several years head start. Moreover, if the current trend of declining cellular prices continues, the PCS operators will face a higher hurdle rate in the future and a lower pricing umbrella under which to discount. These factors were perceived as advantageous for SMR companies because they rely on prices that are similar to cellular prices rather than a big discount from cellular prices.

Two schools of thought have been offered about the types of products PCS companies will provide in the United States. According to one school, PCS could be a low-priced cellular service, akin to what Mercury offers in the United Kingdom. Mercury offers services similar to existing cellular at a much lower price designed to attract a high volume of business, particularly in the consumer market.

According to the second school of thought, PCS will be an advanced cordless product. U.S. consumers currently use 40 million low-quality analog cordless phones. PCS companies could offer the owners of these phones, in return for signing up for PCS service, high-quality digital phones that can be used outside, as well as at home and in an office. Such a product could be very attractive. This approach to the business would be a likely one for the proposed AT&T–McCaw venture because AT&T produces 40 percent of the cordless phones in the country and McCaw Cellular is the largest cellular company. If PCS follows this path, it will compete with the landline networks more than with existing cellular businesses. Such a strategy would also result in incremental sales for the company rather than cannibalization of existing cellular service.

Given the competitive environment, investors interested in PCS must consider two issues. The big issue is elasticity of demand. The early results in England are encouraging; subscribers who are new to mobile communications are being attracted to PCS's lower prices. For example, Mercury One to One, although it offers more a cellular-type system than a PCS system, reported in its first month of operation that about 80 percent of its subscribers did not previously use cellular services. That number is encouraging to potential PCS operators in the United States. The second major issue is that of costs. The question is whether costs can decrease enough to maintain margins at lower prices. Given variable marketing costs, decreasing costs are possible, but PCS margins will be inherently lower than cellular margins because of lower prices. The question will be whether volume will be able to make up for the lower prices and margins.

Paging

Although it is not often considered by analysts following the cellular industry, paging is another major wireless business. The paging industry is growing rapidly as it successfully penetrates the consumer market with units that are smaller than phones. Paging is one of several wireless services that are merging into the cellular companies.

Data Transmission

An entirely new type of business for the cellular companies is the transmission of data over cellular frequencies. IBM, in conjunction with McCaw, GTE Corporation, and several cellular carriers, has invented a packet switching technique called "cellular digital packet data" that can deliver data over cellular frequencies. The technique piggybacks on existing infrastructure. The sites, towers, overhead, and staff are already in place, and the incremental capital requirement is relatively small. Thus, cellular companies can charge much lower prices than such freestanding data-transmission companies as RAM Mobile Data or ARDIS, which are joint ventures of, respectively, BellSouth Corporation/RAM Corporation and Motorola/IBM.

Data transmission and information services will be a big business from both a consumer and a business standpoint. Services such as telemetry that allow vending machines to use radio signals to measure when they are out of certain goods or need servicing could be very important to the industry. Their effects are not in the forecasts, however, because no one knows how to quantify them. Some estimates are that data transmission might contribute 15–30 percent of cellular revenues by the end of the decade. Given that the flow-throughs are incremental, that contribution would be a significant factor.

Valuation Techniques

Valuation techniques applied to the cellular companies rely on discounted cash flows of likely long-term performance. The initial valuation technique was on a population, or per pop, basis. This technique literally translated everything into terms of the population of the area served. To determine the value of a cellular system, analysts forecasted out a number of years, accounted for interim cash flows and free cash flow after capital expenditures, and gave a terminal multiple to the operating cash flow before depreciation, interest, and taxes in the terminal year. To find a per pop value, analysts would divide the sum of the present value of the interim cash flow stream and the terminal value by the population of the franchise area.

Valuation began with this approach primarily because of the way the licensing structure began. Many of the licenses were held by four or five different parties; one owner might hold 10 percent of one market and 40 percent of another. Many of the amalgamations that occurred resembled a monopoly game. The ownership interests were like trading chips, and "per pop" was the currency used in the trading.

Per pop values have escalated enormously. Some of the early deals by the phone companies involved bids as low as $10 a pop. When Southwestern Bell bought Metromedia's cellular systems, including licenses for the cities of Chicago, Boston, Washington, D.C., and Baltimore, it paid what was then an outstanding price of $40 a pop. Today most cellular companies are valued at about $150 per pop in the market, although AT&T proposes to buy McCaw for close to $300 a pop. The escalation was caused not only by the demand for systems and the cash that the phone companies had but also by the performance of the industry, which has led to forecasts being constantly revised upward.

Per pop analysis is simplistic and considers all

markets the same, but many differences exist between a pop in New York and one in, say, Dubuque. Rates and usage are much higher in New York because of longer commuting times and so forth. Nevertheless, most cellular companies do not have reported earnings that would allow use of a conventional earnings measure. They have matured to the point, however, of having significant operating cash flow (cash flow before depreciation, interest, and taxes). Operating cash flow multiples are used for analysis because most companies, as a result of historical losses, do not yet pay taxes.

At one share of AT&T for each share of McCaw Cellular (more if AT&T stock is under $53), AT&T would be paying approximately 17 times estimated 1994 operating cash flow for McCaw. That sounds like a high multiple, but based on discounted cash flow, McCaw is well worth the price; McCaw's cash flow has been growing at 30 percent a year. Also, buying the biggest cellular company in the country probably requires an additional premium.

Analyzing operating cash flow is not without problems. Some companies trade at high cash flow multiples—40 or 50—because of where they are in the development cycle, not because of how they well they are performing. The ESMR sector, because it is just starting, does not yet lend itself to cash flow analysis.

Internationally, depending on the market, analysts sometimes use per pop shortcuts. Vodafone uses the concept of equivalent pops. It translates its cellular interests in other countries into U.K.-equivalent pops based on the GDP of each country relative to U.K. GDP. Analysts can do the same for PacTel in the United States. Ultimately, however, the analysis must look at long-term discounted cash flow.

The first factor to consider in long-term discounted cash flow analysis is penetration. Although the estimate for the past couple of years has been 15 percent by the year 2000, a range of 20–25 percent adjusts for expansion of the market through competition.

In addition to determining penetration, analysts must determine which sectors and participants will gain share. Existing operators, including whatever PCS operations they can offer, could retain 80 percent of the business in the year 2000. PCS and SMR systems, both of which will be relatively new, may have 10 percent each. That mix might shift in subsequent years if PCS increases its percentage of the business.

The next factor to consider is revenue per subscriber per month, which has declined from $100 to $70 a month since 1989. The changing customer mix is responsible for some of this decline, and competition will cause an even further decline. Depending on the mix of customers, subscriber revenues should be about $55 a month in the year 2000. (This estimate is not very different from the average revenue per access line for the telephone industry based on a mix of low-end residential users and high-end business customers.)

The established companies have gross margins in the low 70 percent range with operating cash flow margins of 45 percent. Those margins will decline, however, as revenues decline. Increased volume may make up for some of the decline in terms of absolute profit, but with lower prices, margins might be in the mid-30s by the end of the decade.

Determining the appropriate terminal value and discount rate is critical. These figures are subject to interpretation and make a significant difference in valuations. I use 12 to 13 times operating cash flow for the terminal multiple. The multiple is higher than the terminal multiple I use to value cable or broadcast companies (about 9 to 11 times) because the growth rate reflected in the predicted cellular penetration, the low 20s by the year 2000, will be well above average.

The discount rate is a function of interest rates and risks. Existing cellular operators will have the lowest discount rates because they have the lowest risk. They have completed the construction cycle, have capital, are currently profitable, and enjoy relative certainty of demand. The newer services warrant higher discount rates, which is another reason the per pop values are lower for some of the ESMR companies. The cellular business is initially capital intensive but becomes less so in time because the incremental costs, particularly with digital technology, are fairly low and the industry can quickly generate a lot of free cash flow.

Investment Considerations

Over the years, the values and stock performance of cellular investments have changed drastically. Following the initial burst of enthusiasm over the tremendous success of cellular technology, the stocks of cellular companies came down to earth as the specter of competition from PCS and other technologies rose. Despite the industry's strong performance, market conditions for cellular stocks in 1989 and 1990 were difficult.

Following a hiatus of several years, cellular stocks performed well in 1993. The group was up 41 percent through early November, compared with 4 percent for the S&P 500 Index. ESMR companies performed even better. Their stocks more than doubled in 1993 as Nextel's acquisitions gave credibility to the evolving industry.

International business can be a large element of evaluating the opportunities and risks in cellular companies' futures. Many domestic and foreign companies traded publicly in the United States have a substantial international presence. Two important examples are Vodafone, the large nonwireline operator in the United Kingdom that is traded in the United States, and Pacific Telesis Group, which is planning to sell an interest in its cellular unit to the public and then spin the unit off entirely after a six-month seasoning period. PacTel, the unit being spun off as Air Touch, has more pops outside the United States than inside. Its holdings include major interests in cellular networks in Germany, Japan, Portugal, and several other countries.

Institutional investors with size constraints have limited choices among cellular stocks. McCaw, the largest company in the cellular industry, may be sold to AT&T. As a result, the cellular spin-off from Pacific Telesis, will become the largest publicly traded company in the cellular industry. It will be the only remaining big-capitalization domestic cellular stock for consideration by institutional investors constrained by size. It is slightly smaller than McCaw in terms of market capitalization but has a large international business.

Two other companies are subject to ultimate takeout agreements. Under an agreement following a battle with BellSouth a few years ago, McCaw owns 52 percent of LIN Broadcasting Corporation. The agreement calls for an appraisal in early 1995 as to the private market value of LIN. The private value is what an arms-length buyer would pay for the whole company as opposed to what its stock is selling for in the stock market. McCaw, which by then could be part of AT&T, has the right to buy the rest of LIN at the appraised price; otherwise, it must offer to sell its share with the rest of the company to the highest bidder. The buyout value should be at least $135 a share; the stock currently trades at about $112. Investors have to determine whether the rate of return between now and mid-1995 is sufficient in light of expected interest rates and alternative investments.

Cellular Communications, which originally had the licenses for most of Ohio, has a stepped buyout agreement with PacTel Corporation. In 1991, PacTel bought 5 percent of Cellular Communications and combined that interest with its properties in Michigan, including Detroit. The resulting joint venture operates in both the Ohio and Michigan markets. PacTel has bought additional shares of Cellular Communications in the open market to achieve 12 percent ownership. It has the right to increase ownership to 27.5 percent before October 1995, after which PacTel may make a tender offer for about 23 percent of the Cellular Communications stock. As with the LIN Broadcasting agreement, the Cellular Communications agreement calls for an appraisal process in 1996, when PacTel could buy the rest of Cellular Communications or sell its share in the open market. PacTel seems certain to buy out the rest of the company, but as with the LIN deal, the investment decision depends on the investor's required rate of return.

Several smaller publicly traded companies are also available for investment. The biggest of the small is Vanguard Cellular Systems, which is often considered the next most likely takeout candidate. Vanguard has about 60 percent of its assets in systems in Pennsylvania from Harrisburg to Philadelphia (but not Philadelphia), which the company calls its supersystem. The rest are scattered in other clusters.

Other than the PacTel spin-off, most publicly traded telephone companies do not offer enough participation from their cellular operations to make much impact on company stock performance. The exceptions are Telephone & Data Systems (TDS) and Century Telephone. TDS, an independent telephone company, owns approximately 80 percent of U.S. Cellular. The majority of Century Telephone's market value is attributable to its cellular interests rather than its telephone interests.

Other, smaller companies have scattered clusters of operations in medium-sized cities across the country. Centennial Cellular is a spin-off of Century Communications in a joint venture with Citizens Utilities. Cellular Communications of Puerto Rico, a spin-off of Cellular Communications, Inc., has the nonwireline license for that island. These companies are in the early stages of development, and they are growing faster than the established companies. As a result, they were among the better stock performers in 1993.

Cellular investors can also choose among foreign companies. Vodafone has U.K. licenses and almost as many pops in other countries. Rogers Cantel is the national nonwireline carrier for Canada. Millicom Cellular International, technically a U.S. company with a portfolio of minority cellular interests, has cellular interests in a number of Third World countries. The cellular business is promising in lesser developed countries because it offers a substitute for the landline telephone. In many of these countries, conventional telephone service is limited and landline phones are scarce. For businesses and foreigners working in these countries, the cellular phone is often the main source of communication.

Hybrid companies function in a variety of businesses and offer a variety of services. ESMR companies are one example. Another would be Associated Communications, which owns a portfolio that includes a large position in Tele-Communications, Inc., and cellular interests in Albany, Buffalo, Rochester, and Pittsburgh. Associated just agreed to sell its domestic cellular portfolio to Southwestern Bell.

Conclusion

Cellular penetration is still a fraction of what it is likely to be by the end of the decade, even though it has risen dramatically. Investment in cellular operators still looks very promising even though they will meet more competition, because the offset to the competition is a bigger marketplace, and we believe the current operators will get the largest share of it.

Question and Answer Session

Dennis H. Leibowitz

Question: What is a viable strategy for a cellular operator in light of the upcoming PCS auction? Will the so-called winner's curse be a problem?[1] What effect will PCS licenses have on rural service areas and on metropolitan statistical areas?

Leibowitz: I do not know the best strategy. Operators are allowed to apply simultaneously for multiple markets as part of an overall sealed bid. In other words, they could bid for ten regions in one bid. If the bid price they offer is higher than the cumulative bid of competitors in all the individual markets, they win.

This FCC invention will probably drive up prices. The loudest complaints about the process came from PCS applicants. The cellular industry trade association thought the FCC did a magnificent job, so that indicates what the economics are.

Because PCS requires many cells and low prices, it probably won't be much of a factor in the small markets. The economics do not justify it. In the bigger markets, companies will amalgamate frequencies. Companies with two cellular and one SMR license, will probably acquire two or three, certainly not seven, PCS licenses.

The most valuable licenses are the 30-MHz, broad-spectrum ones, but that group is also where the bidding is likely to be the highest. For additional spectrum, existing operators will certainly apply for the 10-MHz licenses on which they are entitled to bid. A winner's curse is certainly a risk because of the auction process.

Question: What are the economics of overlaying PCN on the cable infrastructure?

Leibowitz: Even with a lot more lower priced cells, the total system costs of PCN are not that different from cellular, but one way to make PCN economics more favorable is to use another company's infrastructure. The cable industry's infrastructure is well suited to PCS—if not as a licensee, at least as a distributor. The cost of transporting calls from hundreds of cells all over the place can be astronomical when landline or microwave services are used. Piggybacking off the cable system entails putting antennas on telephone poles and connecting them into the cable plant. Cable companies can do the cellular signal processing at the head end of the cable company and amortize it over more cells. As a result of these economies, cable operators will be a major factor in the application process for PCS licenses.

Because access to infrastructure will be a major factor in PCS applications, cable companies will be bidders, particularly wherever they have cable systems. TCI has indicated that it will bid. Time Warner, which has been very aggressive in this area, will also bid. If they lose the application process, such companies may rent their facilities to the winners to collect extra revenues on their existing capital investment. Companies without existing infrastructure will be at a competitive disadvantage in the PCS market.

Question: What are the competitive implications of the effort to start a nationwide, seamless network, and what could go wrong with that development?

Leibowitz: Companies that offer the ability to use one phone anywhere in the country for call forwarding, data delivery, and paging have a tremendous marketing advantage over those that do not offer a similar package of services. In particular, a single vendor offering a single bill with volume discount pricing on a variety of services will have a competitive advantage with respect to corporate business, which is the most profitable segment of the market. So, cellular operators will want to put such packages together.

The main competition will come from the SMR sector. The SMR companies are spending billions of dollars to build systems. They have raised the ante significantly above what was required by the original cellular approach of building out six major cities one by one.

Question: What drives the forecast for growth in penetration in cellular?

Leibowitz: The combination of lower prices and word of mouth explains why cellular is doing so well. Not only are customer prices coming down, but companies are offering a price plan for everybody's needs. People who have mobile phones will not give them up, and word of that attitude toward the product is spreading.

Question: Please elaborate on the changes in the average monthly bill from $100 to $70.

---

[1] The "winner's curse" applies to a situation in which winning is worse than losing because of, for example, the up-front costs of winning.

Leibowitz: The change in the average monthly bill is a function of mix. Many companies that have analyzed the change found that, other than during the recessionary period, existing subscribers are not using the service any less. The real reason for the decrease in the average is that new subscribers to the service are lighter users than previous subscribers. At the same time, the absolute retail price is increasing. By offering bundled minute plans, companies have raised the floor. They experimented with rates like $9.95 a month and $1.00 a minute and found that many people were buying the service for emergencies and security but never used the phones at all. At one point, 10 percent of Rogers Cantel customers did not have any billable minutes.

Question: How does the U.S. cellular industry compare with the industry in the rest of the world? Are we moving toward universal global standards in SMR or cellular?

Leibowitz: The U.S. cellular industry is doing better than other nations—except, as I discussed, Scandinavia—because of the competitive environment here, the pricing situation, the relative U.S. income, and the fact that the U.S. industry is more attuned to changes in communication technology and more readily adaptable than more conservative countries in Europe. Many of those countries have only recently licensed competition and may do as well as the U.S. industry when competition increases.

Mobile satellite systems, such as those offered by American Mobile Satellite Corporation, can offer the equivalent of cellular services to people in remote areas. Motorola's Iridium satellite system will do the same on a worldwide basis, although the handsets will be expensive because they must be able to transmit as well as receive.

The industry is not moving to universal global standards. The PCS standard in Europe is universal for Europe, but it is different in the United States. Japan offers different services on altogether different frequencies. The industry did not adopt universal global standards for analog technology, and I doubt that the industry will adopt universal standards for digital.

Question: How do you factor debt into a free cash flow analysis of cellular companies? Does the definition of free cash flow include interest payments?

Leibowitz: Free cash flow is net income plus depreciation and other noncash charges after capital expenditures and dividends. It would be after interest payments, which is where debt would be reflected.

Question: What is the relationship between depreciation and capital expenditures? What is the reinvestment requirement?

Leibowitz: Depreciation is a function of capital expenditures, and it is based on gross plant. The average depreciation life of cell site and other infrastructure is about ten years.

Question: Do the terms "30 MHz" and "20 MHz" refer to the bandwidth or broadcast frequency?

Leibowitz: They refer to the amount of spectrum or bandwidth in the frequency assigned (800 MHz is the frequency for cellular; 1.8–2.1 GHz for PCS).

Question: Are any new technologies that will raise the forecasted penetration rate likely—for example, the new lightweight battery by Valence Corporation.

Leibowitz: Not in the foreseeable future that I know of. This would be an enhancement of the existing products.

Question: Will cellular complement wireline, fully digital, broadband service or have the capability to compete directly against wireline, especially in the home?

Leibowitz: Both, and they are likely be a part of a communications portfolio of offerings by integrated companies, particularly the RBOCs.

Question: In the move from analog to digital cellular, will cellular companies be able to operate on existing infrastructure so that the consumer will only have to invest in a new phone?

Leibowitz: Phones will be dual mode (analog/digital) and will default to analog wherever digital of the same type (TDMA, CDMA) is not available. Cellular companies will change radio channels; the rest is the same whether analog or digital (because the switch is already digital in most cases).

Question: Could you bring us up-to-date on cellular digital packet data (CDPD)?

Leibowitz: It is being implemented by several carriers now, starting with McCaw Cellular in Las Vegas, and it should be widely available by year-end.

Question: Canada and several other countries have recently licensed personal cordless telephone systems or two-way CT2 technology. What will be the future of this technology in the

United States and elsewhere?

Leibowitz: It has not been successful in the United Kingdom, where it was one way, but it has met with some success in Hong Kong and Singapore. A PCS system is likely to be more successful.

Question: What could go wrong with SMR? Can cellular providers do anything to slow or stop its development?

Leibowitz: The technology is still new and it could be that ESMR operators will have to cut prices to induce cellular subscribers or prospects to use ESMR instead.

Question: What is the outlook for SMR in those countries that have national cellular systems?

Leibowitz: SMR is not a factor in competition with cellular for regulatory reasons, except possibly in Canada, where an ESMR structure is being considered.

Question: Do the enhanced services offered by ESMR (data transmission, dispatch) and the higher terminal costs ($1,000–$3,000) attract a unique customer profile?

Leibowitz: The main unique appeal would be to those subscribers who are both dispatch and cellular subscribers.

Question: How will PCS effect rural services areas? Which poses the greatest threat to cellular providers, ESMR (Nextel) or PCS providers?

Leibowitz: Rural areas are not likely to be viable for PCS but ESMR might be. On the other hand, PCS is more of a threat to urban cellular than ESMR because of price.

Question: In light of the fact that several RBOCs are buying CDMA technology, is the industry moving toward this technology?

Leibowitz: The majority are implementing TDMA, but CDMA could work out for later generations of digital.

Question: How long do cellular companies typically follow policies using loss leaders and giving away equipment in new markets?

Leibowitz: It depends on the competition in each market.

Question: Spectrum is widely perceived to be public property. Are cellular profits that depend on enormous margins vulnerable to expropriation by cash-poor federal, state, or local governments?

Leibowitz: In the sense that auctions drive up prices, yes, but this only applies to new carriers.

Question: Will PCS require medical testing to determine the risks of exposure to higher frequencies?

Leibowitz: No, the power is much lower than amounts that would pose a health risk.

The Telecommunications Equipment Industry

Maria F. Lewis, CFA
Managing Director
Cowen & Company

> The telecommunications equipment industry is enjoying rapid growth as the cable television and telephone companies work to bring interactivity to the home. A favorable regulatory environment, new technologies, and new markets are also fueling growth.

The telecommunications equipment industry is a dynamic sector with abundant growth and investment opportunities worthy of further investigation before some timely capital commitments are considered. This presentation focuses on telecom equipment manufacturers and vendors that supply infrastructure equipment to local and interexchange carriers (a $70 billion to $80 billion business); cellular and personal communication service carriers ($8 billion); and cable television multisystem operators and other entertainment providers ($2–3 billion). In total, the world market in these sectors approaches $100 billion, and it is growing rapidly.

Industry Structure

The telecommunications equipment industry is divided into a number of product-market segments. The major categories of equipment sold to carriers or service operators are central office switches, transmission and intelligent network (IN) equipment, wireless infrastructure, and cable TV distribution plant and head-end gear. This discussion of video distribution equipment will focus principally on cable TV equipment, not satellite gear, consumer earth stations, or over-the-air broadcast systems. Other product segments include business communications equipment and customers' premises equipment (CPE); data-networking equipment (the adapters, hubs, servers, and routers that make up the infrastructure of network computing); and terminals or handsets (corded, cordless, cellular, or the new class of personal digital assistants).

Key players can be classified by market segment and product-line scope. As shown in **Table 1**, they include major vendors with full product lines, as well as niche suppliers with narrow product lines selling to local telcos, IXCs, and international postal, telephone, and telegraph carriers (PTTs). The major vendors have familiar names, but the niche providers are all Nasdaq listed and are probably less well known. As seen in the table, even niche companies can be large firms, as in the case of Harris Corporation and GM Hughes, which participate in the wireless sector (cellular and PCS carriers).

The leading wireless equipment vendors overlap with suppliers to local exchange carriers, PTTs, and regional cellular operators. The CATV suppliers differ considerably from the suppliers to other sectors. Until recently, the cable industry has been treated as a distinct subsector with different technological, financial, and procedural considerations, but the lines between cable operators and telcos are poised to blur substantially, and the vendor lists serving each constituency will overlap considerably more in the future. AT&T has already begun aggressive marketing to the cable industry, and it may soon be joined by L.M. Ericsson, Northern Telecom, and Newbridge Networks.

Newer market participants have gained visibility from recent initial public offerings (IPOs). The industry's robust growth and heady valuations (at least by historical standards) have led to a flurry—some would say blizzard—of IPO activity. The latest wave brought to market an interesting mix of yet-to-be-profitable "concept stocks" such as Broadband Technologies, First Pacific Networks, and Qualcomm, which went public on the perceived strengths of their future market prospects and current product or technology positioning. In addition, both Qualcomm and First Pacific completed successful follow-on, secondary offerings, although the companies are

Table 1. Major Telecommunication Equipment Suppliers, by Sector

Sector	Full Product Lines	Niche Providers
Local and interexchange equipment	Alcatel AT&T Ericsson Northern Telecom Siemens	Andrew Corporation ADC Telecommunications DSC Communications Newbridge Networks Tellabs
Cellular and PCS equipment	AT&T Ericsson Motorola Nokia Northern Telecom	ADC Telecommunications The Allen Group Andrew Corporation Harris Corporation GM Hughes
CATV MSO equipment	General Instrument Scientific–Atlanta	AT&T C-Cor Electronics Philips Broadband Pioneer

Source: Cowen & Co.

still very much in the red.

Market Forces and Participants

Market forces fueling the industry's growth include catalysts to increased customer demand; regulatory initiatives and outcomes; new carriers, services, and technologies; and the market opportunities presented by lesser developed countries that are building infrastructure.

Customer Demand

Customer demand for telecommunications equipment and services is strong. Demand for new services is fueled by a heightened desire for business competitiveness, greater communications mobility, and increased personal productivity through modern communications and data-processing tools, as well as greater access to and diversity in entertainment and information sources. Competitiveness is shaped by the globalization of many industries, which together with a distinct shift in data-processing architectures from hierarchical mainframe-based computing to personal- and workstation-based computing, heightens the need for responsive communications equipment and services. Consumer expectations regarding information, shopping, and entertainment services also have grown more sophisticated. As evidenced by paging and cellular subscriptions, businesses and consumers have embraced voice and data mobility with a vengeance.

Regulation

Regulation has historically played an important role in carriers' business development and, by inference, in equipment suppliers' opportunities. Today, deregulation is an enabling factor helping to fuel the strong growth that communications infrastructure equipment suppliers now enjoy. Industry distinctions are blurring, and new operators are being formed.

Regulatory decisions also are having a positive effect on equipment spending. The FCC's video dial tone decision, for example, which ruled that carriers could carry a video data stream much like a phone call, spurred a number of market trials involving new technologies and new suppliers. Cable reregulation, by strictly regulating basic service, creates incentives for acquiring advanced services and the hardware to deploy them. The same is true of Bell Atlantic's court victory granting it the right to invest in cable content, even though this ruling does not automatically apply to other regional telephone companies. Finally, the FCC's spectrum-allocation process promises to create as many as seven new entrants per market, each needing wireless infrastructure and each of their customers needing portable terminals. As always, once liberalization is set in motion by choice or by chance, regulators are helpless to constrain the powerful forces of advancing technology, declining costs, and rising consumer expectations.

Although not dating from a single event or decree, the reasons carriers acquire equipment and software have changed. What used to be driven solely by the rate base and regulated rate of return is now influenced by a desire to stimulate revenue growth, contain costs, and enhance earnings. Slow subscriber growth and the declining costs of technology brought about this change, and now the companies find the only effective formula for increasing earnings is via efficiency improvements, flexible return-sharing regulation, and the successful introduction and marketing of value-added services.

Privatization and New Global Markets

Internationally, PTTs are privatizing. Governments are selling interests to the public or to consortia of commercial companies bringing fresh capital and often new construction mandates to their regions. Some 10–15 countries, including New Zealand, Australia, Hungary, Mexico, and several other Latin American nations, have already privatized their national telco carriers. Industrial giants like Germany, France, and Japan have further liberalization and/or additional public equity tranches planned. Some 30–40 additional privatizations are expected by the end of the decade. Whether in a developed or developing economy, privatizations often provide significant infusions of capital. A good portion of this capital, as well as operations expertise, has come from the RBOCs, which sought to diversify after the AT&T divestiture ruling. Privatization stimulates—indeed, usually mandates—increased construction spending in response to what is often the first application of the profit motive for a formerly nationalized carrier.

Another key catalyst to global infrastructure demand has been new markets. As highlighted in **Figure 1**, the density of the world's phones varies greatly between the haves and the have-nots. This has long been the case, although the latter's access to capital and appetite for entering the information age are changing the situation. China is at the bottom of the list, but in several years, it expects to install 20 million new central office lines a year, compared with 10 million in the United States. China will prove more attractive to equipment producers in a number of other ways, including design and construction potential, as well as favorable profit characteristics. Eastern Europe and the Commonwealth of Independent States (CIS) may soon follow suit. The rebuilding of East Germany is already well underway, although largely to the exclusive benefit of Siemens and other German suppliers.

The developing countries have several other common characteristics. Because very little existing infrastructure is in place, in many cases, they can leap-frog to more cost-effective and innovative solutions such as wireless communications. They present an opportunity to plan and build modern networks almost literally from the ground up. GM Hughes is currently delivering on an approximately $100 million award in the CIS whereby its wireless fixed-access local-loop technology is delivering basic fixed subscriber services at $1,000 per connection; the same services would cost some $3,000 per connection if cabled in a traditional star architecture.

New Carriers, Services, and Technologies

Another factor that has added impetus to the growth in telecom construction spending has been the emergence of new carriers challenging the industry's established monopolies, duopolies, and oligopolies. They range from competitive access providers, such as Metropolitan Fiber Systems, to new long-distance and cellular or cellular-like carriers. The most prominent of the latter category is Nextel, but other would-be providers include Geotek Industries and American Mobile Satellite Corporation. Such developments are no longer distinctly American; multiple operators, particularly of the wireless variety, are active in a number of nations.

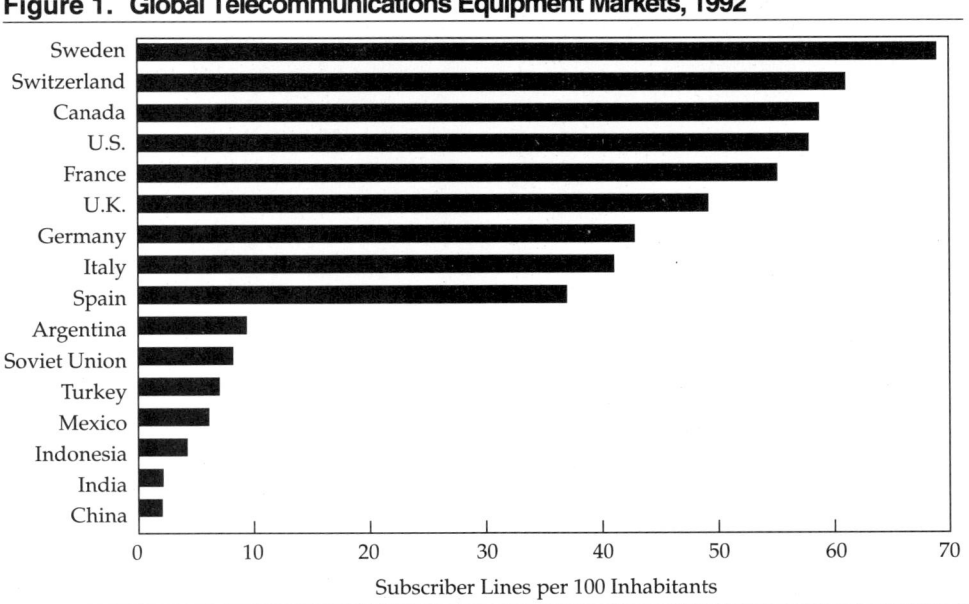

Figure 1. Global Telecommunications Equipment Markets, 1992

Source: The World's Telephones, LM Ericsson, 1992.

Several data-only mobile carriers have emerged to challenge cellular carriers. They are also spurring paging operators to provide increased functionality. New value-added service providers have emerged to take advantage of open-network-architecture rules and advanced network and customer premises equipment capabilities, specifically in fax, messaging, and other outsource services.

Cable operators are no longer the only players in the entertainment delivery game. Other firms are joining in through initiatives like Hughes' DirecTv, a direct broadcast satellite service, and Cellular Vision's millimeter wave over-the-air programming distribution. Although few of these companies offer investors direct participation in their growth (the exception is paging), they all generate increased telecom equipment demand, making equipment producer stocks an attractive way to play these emerging market segments.

The new carriers have prompted new services deployment. Whether provided by existing or new entrants, expanded telecommunications capabilities require increased equipment spending. The spending can range from a relatively modest incremental investment in new processing or database hardware and software to a completely new network infrastructure, as in the case of Hughes' DBS initiative and Nextel's enhanced specialized mobile radio services. Sometimes the investment is in an overlay system such as virtual private network business services or single-number capabilities. International carriers frequently build overlay systems because they are the quickest way to get competitive offerings into the market. Other carriers are creating distinctive services through marketing and perhaps accelerated evolutionary investment, as may be the case with cellular carriers' pursuit of PCS. Sometimes, however, the only way to provide new services will be to build from the ground up, or the sky down, as the case may be.

New carriers and new services have been spurred in large part by new technologies. Technologies that have played a major role in the acceleration of the industry's development include digital switching, transport, processing, control, and compression; fiber optics and broadband wireless capability; and the rapid diffusion of intelligent network capabilities. The introduction of new technologies, by making service- or cost-based competition feasible or simply economically viable, may launch a replacement cycle by encouraging early retirement of existing equipment and/or new entrants into the industry.

Product/Market Sectors

The telecommunications equipment market can be classified according to product end uses. The major subdivisions are central office switching equipment, transmission and IN components, wireless infrastructure, and CATV equipment.

Central Office Switches

For a producer to lack a presence in central office switches once meant virtual exclusion from the equipment market. Although that is no longer the case, these "neighborhood switchboards" remain the cornerstone of full-line suppliers' product lines. With growth of less than 10 percent annually, however, the switches represent a maturing business in most developed countries. Elsewhere, demand is strong, growing more than 20 percent annually, and all areas have a healthy appetite for software-intensive advanced capabilities. Those capabilities, although sometimes delivered on other platforms, are still very much the province of central office switch suppliers, tempting them to price switches like low-cost razors, assured they will have a long and healthy market for high-margined software applications.

Central office market opportunities vary by customer geographic base, as seen in **Table 2**. Not long ago, these suppliers served only traditional telco customers around the globe. Now, competition exists almost everywhere in a variety of forms such as IXCs, cable, CAPs, and wireless. In general, suppliers to non-North American markets face firmer demand, as do those who have penetrated nontraditional operators. No region or customer constituency is universally unattractive. Wireless, however, presents an

Table 2. Central Office Switching Opportunity, by Geographic Customer Base

Type of Supplier	North America	Latin America	Western Europe	Eastern Europe	Pacific Rim
Traditional telco (RBOC, PTT)	L	H	M	H	H
IXC	L	L	M	L	M
Cable TV to telco	M	L	M	M	L
New entrants (CAPs)	H	M	M	L	H
Cellular/wireless	H	H	H	H	H

Source: Cowen & Co.
Note: H = high, M = moderate, and L = low.

excellent opportunity regardless of location.

The top suppliers' geographic presence also varies, as seen in **Table 3**. The European suppliers Alcatel, Ericsson, and Siemens have historically dominated the global market, but they have been challenged recently in Latin America by AT&T and in the Pacific Rim by Northern Telecom. Although the big four in Japan (NEC, Toshiba, Fujitso, Matsushita), represented by NEC, have been a big factor in their own market, the Japanese have been singularly unsuccessful outside the Pacific Rim. High software and service intensity, as well as rampant nationalism, help explain the absence of much Japanese success outside their home market.

lay and establish the parameters of opportunity for all suppliers. Suppliers from industries such as computers and software would be wise to find partners that can help them gain entry to ATM, video servers and compression, interactive cable, and PCS.

IN nodes lend new signaling, data base, messaging, and management capabilities to previously "dumb" or less flexible networks. IN nodes, for example, are responsible for translating 800 numbers to phones that ring or for providing single-number service to cellular/PCS/residential subscribers. At one time, virtually all network intelligence was central office switch resident, but now IN nodes have blossomed throughout the network, particularly

Table 3. Geographic Presence of Central Office Switching Companies

Company	North America	Latin America	Western Europe	Eastern Europe	Pacific Rim
Alcatel		√	√	√	√
AT&T	√	√			
Ericsson		√	√	√	√
NEC					√
Northern Telecom	√	√			√
Siemens	√		√	√	

Source: Cowen & Co.

Transmission and IN Components

Transmission and IN components have overtaken central office switches to constitute the industry's single largest sector; they now account for 30–40 percent of industry volume. These markets enjoy better growth characteristics than other sectors as broadband intelligent networking gains acceptance throughout the world in sophisticated voice, data, and video applications. Transmission is the bottleneck in many networks. Switching was digitized first, and now transmission is poised to follow, which will increase the capacity, flexibility, and manageability of this process. The architectural elements of the two operations are so similar that discerning where switching systems end and transmission systems begin may soon be difficult.

Transmission and IN trends will benefit telecom equipment suppliers. Cable and telco plans to pave the electronic highway to the home mean rapidly expanding sales of fiber, optoelectronics, and broadband radio equipment. Similarly nontraditional and existing suppliers will find opportunity in distinctly new network segments such as video servers and video compression and decompression gear. Even though the definition of transmission and IN equipment is expanding to include asynchronous transfer mode (ATM), interactive cable, and PCS, telephony elements remain the predominant architectural over-

close to the end user, in a process not unlike the distribution of data processing resources from mainframes to PCs. Because IN nodes are acquired as independent network elements, full-line suppliers face the increased costs of marketing an additional sale, but the pay-off is participation in a high-margin growth opportunity serving carriers and their customers alike.

Wireless Infrastructure

Although the wireless sector is small at 10 percent of equipment spending, its growth is accelerating faster than that of the other sectors. New subscribers to existing networks and new networks for new services such as PCS and mobile data are the predominant catalysts to continuing growth. As with broadband network builds, telephony elements and functionality have superseded radio competency as key components of wireless networks. Wireless growth is currently being fueled by a broad and diverse variety of factors: replacement spending as existing networks move from analog to digital technology; new networks of newly licensed carriers; new users of previously licensed spectrum, as in the case of the specialized mobile radio industry; and use of wireless, for cost and expediency reasons, as a substitute for wired local loops, especially in developing markets.

During the past century, the global telephone industry has grown to its current level of approximately 500 million subscribers. That number may double within the decade with more than one out of five (in some countries one out of two or three) new subscribers coming via a wireless mode. As shown in **Figure 2**, wireless common-carrier subscriber growth is expected to average at least 30 percent annually through 2000.

CATV Equipment

The $2 billion to $3 billion global cable TV equipment market is poised to rival the 1990s growth prospects of wireless, which are pegged at 20–25 percent annually. Domestically, this market is driven by upgrades and enhancements as operators strive to deliver on promises of many hundreds of channels and interactivity.. Internationally, the requirements are often much more modest—6 over-the-air channels in a recent Saudi Arabian bid, and often only 10 or 12 are needed elsewhere. International orders are attractive, however, because in many cases, an entire infrastructure must be built to distribute and receive the programming, which can lead to sizable, multiyear awards.

Cable operators, like the telcos, are increasingly deploying fiber optics and rapidly migrating from analog to digital technologies. Unlike the telcos, however, cable operators routinely fund subscribers' television set-top gear, a new generation of which will be precipitated by the networks' conversion to digital technology. As in telco circles, alliances are forming rapidly in this sector. Each of the domestic duopolists has an ally for the design and development of next-generation, navigable set tops. Navigable set tops will provide end users with the capability of intelligently honing in on desired selections, rather than channel surfing through 500 airwaves.

Many alliances, such as U S West–Time Warner and the proposed Bell Atlantic Corporation–TCI combination are cross-border, highlighting the growing convergence of telco and cable ambitions. Network plans and plant are taking on an increasingly common look with fiber optics, ATM, and digital video compression playing major roles for each segment. These one-time competitors are also increasingly pairing as allies to defray capital costs and programming access expenses and to leverage complementary realms of experience and expertise. The pairings also mean equipment suppliers should be winners regardless of consumers' ultimate appetite or willingness to pay for advanced services.

Competitive Dynamics

As elsewhere, competition among equipment suppliers is tough. Barriers to entry—substantial R&D and capital requirements and significant volume-based manufacturing efficiencies—are high and rising. Gross margin trends are somewhat conflicting. Margin increases fueled by software intensity and volume growth are eroded, in large part, by competi-

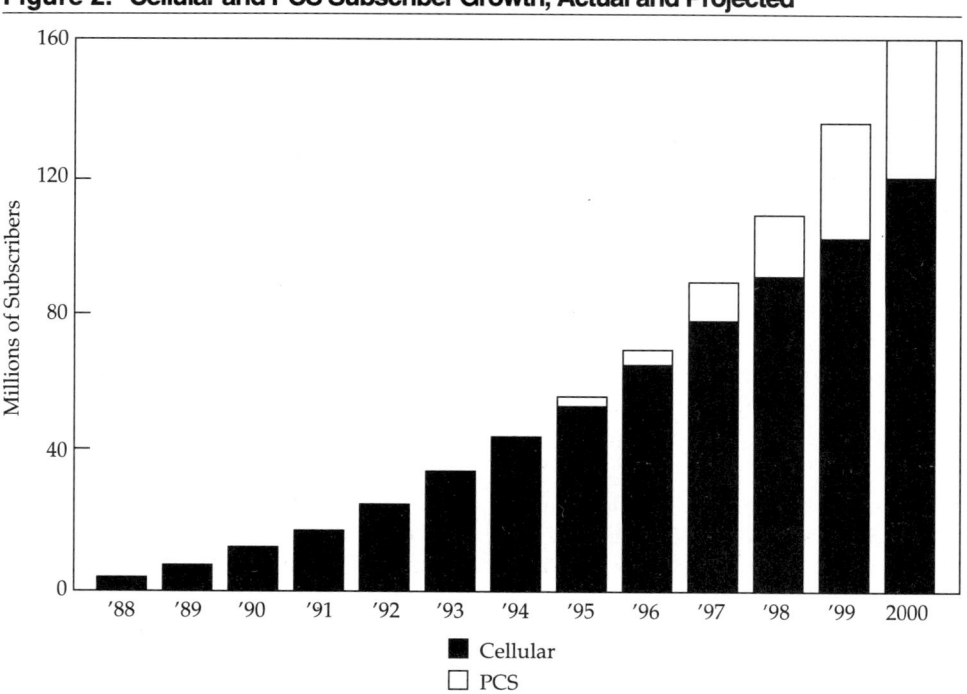

Figure 2. Cellular and PCS Subscriber Growth, Actual and Projected

Sources: Cellular Telecommunications Industry Association (CTIA) actuals and Cowen & Co. estimates.

tive pricing.

The importance of price varies by supplier entrenchment and market maturity within a sector. Price considerations are often secondary to product differentiation; service; support; and in developing countries, financing, manufacturing, and jobs. Central office switching is experiencing heightened price competition in developed and in some developing countries. In the developing countries, this trend is led by privatization investors, especially RBOCs, that know prices in the rest of world. Some cellular markets are also experiencing price competition as producers hope to establish themselves in digital transmission. More mature transmission equipment, such as channel banks or long-distance radio, also compete on price. Price competition does not currently affect sophisticated T–1 (a standard telecommunications channel delivering 1.544 megabits per second of bandwidth) and ATM equipment; SONET standard cross-connects; cable equipment; and aftermarket software, which has few, if any, suppliers other than the original central office manufacturer.

Telecom equipment products are subject to improving performance-to-price ratios and declining price curves; price decreases of 5–8 percent annually should be considered normal. Often, however, as a sector matures, prices will spike downward significantly as established suppliers attempt to win market share from one another. In the U.S. central office market, for example, producers' gross margins have been badly battered as the conversion of switches to digital technology grows complete and AT&T and Northern Telecom wage war over the few remaining multiyear awards.

Upgradability is an important element of competition in this market, especially when depreciable lives can extend beyond ten years and suppliers are often judged on the merits of their long-term architectural approach to the market. Within these architectures, product replacement, at least in software and modular enhancements, is accelerating. Those caught flat-footed (as Northern Telecom recently was with central office software enhancements) could see their businesses deteriorate markedly as a result.

Telecommunications customers demand full-service and near-flawless network availability. No one takes a disruption of phone service lightly. DSC Communications' equipment-induced network outages about two years ago led Congress to convene hearings on the matter. Cable operators should not ignore such demands, because they are increasingly likely to come under similar scrutiny from their customers. Cable operators historically have had it easy; their customers have tolerated service disruptions and quality variations. Customers are likely to be far less tolerant, however, if cable is also handling telephone, PCS, information, and transactions services, as these companies are planning to do.

Financing can be crucially important—not only to struggling new start-ups, but also to cash-strapped countries. Motorola and Northern Telecom provided more than $500 million in financing to Nextel (formerly called Fleet Call Communications). In early November 1993, Motorola agreed to provide $260 million more in a transaction involving the exchange of its SMR properties for a 20 percent Nextel equity interest. Similarly, manufacturing and jobs have long played a role in international vendor selection. Local-content requirements in manufacturing are especially important to lesser developed countries such as China.

To thrive in this environment, companies that want to compete as one of the big boys will need a broad portfolio of competitive systems products, expansive market exposure, financial flexibility, and an efficient cost structure. To dart and weave successfully around the large competitors, niche companies will need high-value-added, competitive products that are early to market. They will also need staying power, defined as the ability to sustain R&D, as well as service and support, through a grueling prove-in process and beyond. Niche suppliers have been known to take three to five years to achieve significant, although not necessarily sustainable, telco market access.

The major participants compete in a variety of product lines, as shown in **Table 4**. In general, the cable and telco producers have yet to overlap broadly in product-line participation. Motorola still stands alone as a successful single-segment wireless supplier. All three single-segment vendors, Motorola, General Instrument, and Scientific–Atlanta, will need to acquire or align with superior telco capability in the not-too-distant future if they are to capitalize on the broader market access potentially available to them—and, if they are to defend successfully against incursions by others on their traditional turf.

Financial and Valuation Metrics

The financial indicators to watch in this industry do not differ greatly from those analysts use for other manufacturing sectors, except perhaps the heightened emphasis on R&D spending. Analysts should monitor revenue growth, gross and operating profitability, ratios of debt to total capital or debt to equity, days receivables, and inventory turnover.

Each indicator can provide useful information about a company and the industry. Revenue growth attests to a company's product and market penetra-

Table 4. Major Participants in Various Product Lines

Company	Central Office Switches	Transmission/ IN	Wireless Infrastructure	CATV Equipment
Alcatel	√	√	√	
AT&T	√	√	√	√
Ericsson	√	√	√	
General Instrument				√
Motorola			√	
Northern Telecom	√	√	√	
Scientific–Atlanta				√
Siemens	√	√		

Source: Cowen & Co.

tion and its cyclicality. Gross margin indicates the degree of price pressure, cost-reduction effort, and product mix. R&D expenditures as a percentage of sales can indicate sustainability of market position. Operating margin is the best indicator of a business's profitability net of financial leverage. The debt-to-capitalization ratio is indicative of a company's financial strength and flexibility. Days sales outstanding is a trend indicator of market activity and customer collections. Inventory turnover can indicate product acceptance, demand trends, and potential problems.

Table 5 is a representative snapshot of these indicators for four major producers. The income statement indicators are three-year averages for each company; the balance sheet indicators are 1992 figures. AT&T's R&D percentage appears low because of its large long-distance revenue base, which is well supported by much lower sustainable research investment than its equipment businesses. Ericsson's high gross margin reflects superior product and geographic mix. Its high days sales figure reflects its geographic and customer diversity. An important exercise would be to follow these indicators on a trend, rather than a static, basis.

Projected revenue growth during the next two years varies according to industry category, as seen in **Figure 3**. All four categories—mainline telecom equipment suppliers, CATV companies (essentially Scientific–Atlanta and General Instrument), niche telco suppliers, and new entrants, including the concept stocks—look strong, but new entrants appear exceptionally so, possibly tripling revenues over this two-year horizon.

As with financial metrics, the valuation variables to consider are no mystery. They include price–earnings, price–book-value, and price–revenue ratios and sustainable growth rate. The price–earnings ratio gauges investors' perceptions of growth opportunities, the risk–earnings trade-off, and the rate of capitalization. Price–earnings ratios for suppliers with earnings are now ranging from 20 to more than 30—high relative to the group's historical norms but probably justified by the acceleration in growth rates that has come with the industry's booming demand conditions. The price–book-value ratio provides a valuation of potential opportunities and risks versus the accounting value of the firm. The price–revenue ratio is an indicator of the perception of potential sales growth. Sustainable growth rate gauges the company's long-term valuation.

Valuations vary widely among the four major

Table 5. Financial Indicators for Selected Telecommunications Equipment Companies

Indicator	AT&T	Ericsson	Motorola	Northern Telecom
P&L indicators (three-year averages)				
Gross margin	39%	49%	37%	39%
R&D percentage	5	15	10	12
Operating margin	7	6	7	8
International sales percentage	25	88	30	33
Balance sheet indicators (1992)				
Debt: capitalization	31	12	20	21
Days sales outstanding	62	123	56	98
Inventory turns	13.7	2.5	6.6	5.6

Sources: Company reports and Cowen & Co. calculations.

Figure 3. Revenue Growth Estimates, by Sector (1992=100)

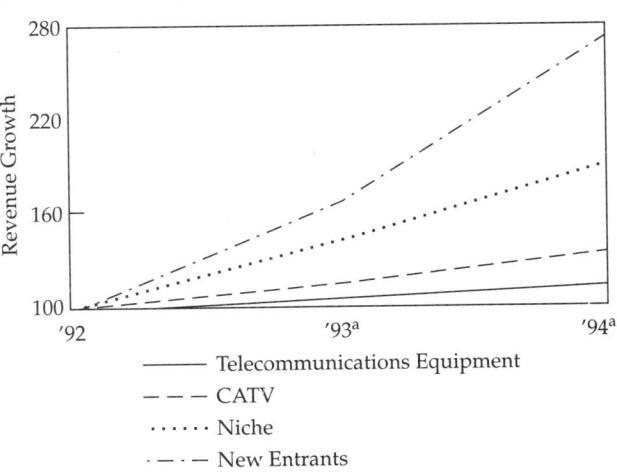

Source: Cowen & Co. estimates.

sectors of the industry, as is apparent from the price–revenue ratios shown in **Figure 4**. The companies without earnings but with perceived superior growth rates trade at the upper end of each scale; those deemed most mature, at the lowest end. Based on recent market results, telecom equipment stocks enjoy high average multiples relative to data-networking shares, even though near-term growth rates are likely higher for the latter.

Conclusion

The telecom equipment industry is big and, by almost any standard, burgeoning. Product and customer definitions are blurring and expanding. Suppliers' business models are adjusting to the industry's new dynamic, meaning gross margins may be lower and operating expenses higher. Hence,

Figure 4. Price–Revenue Comparisons, by Sector

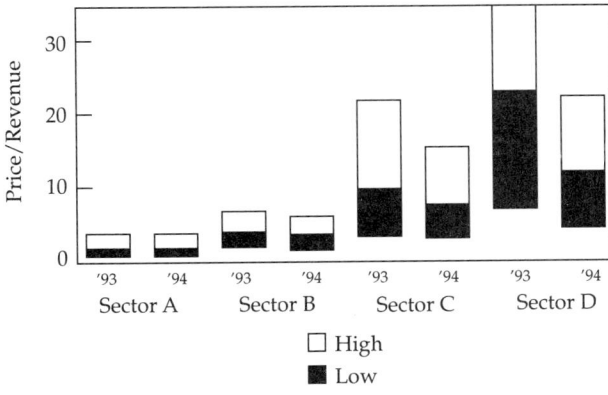

Key: Sector A = Full-Line Telecom Equipment Vendors
  L.M. Ericsson
  Motorola
  AT&T
  Northern Telecom

Sector B = Full-Line CATV Equipment Vendors
  General Instrument
  Scientific–Atlanta

Sector C = Narrow-Line Telecom Equipment Vendors
  Tellabs
  Newbridge Networks
  DSC Communications

Sector D = New Entrants
  Summa Four
  Broadband Technologies
  Qualcomm

Sources: Maria F. Lewis, CFA, based on data from company reports, Cowen & Co. estimates, and stock price data from appropriate exchange.

operating profitability may be lower than historical norms. Earnings growth, however, may well be higher, given the extraordinary confluence of opportunities: strong spending on equipment to meet the strong demand for services, available technology to meet that demand, and a favorable regulatory environment.

Question and Answer Session

Maria F. Lewis, CFA

Question: What are the implications of size differences among the players, particularly the potential fade of small ATM switchmakers and router companies? What risk do they face compared with AT&T and Northern Telecom?

Lewis: Niche companies have various entry points, but companies early to market must provide staying power. What may happen in this industry is similar to what happened in data networking. Router companies, such as Cisco Systems and Wellfleet Communications, were first to establish themselves. They were early to market with high-performance equipment, and they dominated the segment. When International Business Machines Corporation (IBM) and others discovered this market, they were too late to influence the competitive parameters on which the equipment was judged.

Large vendors, such as IBM, Hewlett Packard Company, and Digital Equipment Corporation, have not affected the competitive dynamics of the data-networking market, nor have they captured much of the market except on an original equipment manufacturer basis. The difference between that business and the telephone business is the concentration of buyers. Telephone companies are more conservative, there are fewer of them, and they use fewer suppliers than the high-performance equipment makers. Companies such as Toshiba or Fujitsu will have difficulty entering the market. The Japanese have repeatedly tried to enter the switch market on an ATM-only basis but have been unsuccessful.

Some domestic or local producers such as Newbridge Networks of Canada have worked hard for the past several years to win the trust, endorsement, and Bellcore approvals necessary of these customers. Newbridge is bringing its bandwidth management equipment into more than 100 telcos nationally and will likely be successful in getting its ATM equipment installed at major telcos as well.

Some companies have participated successfully on a niche basis. As customer distinctions blur, there will be more successful niche companies, but the process is easier when the buyers are more diverse.

Question: Please provide an overview of the various switching technologies in terms of demand for interactive services. Who is in the best position to exploit that?

Lewis: Interactivity has two divergent starting points, and the solution for providing interactivity depends on the starting point. The telcos have a star architecture. It is already interactive, but the bandwidth of the channels is very limited. Cable has a traditional trunk and branch architecture. The infrastructure is already wideband. The channels have a lot of capacity but little interactivity. What interactivity exists is asymmetrical: big, wide channels enter the home, but a very small data channel goes back to the programmer or cable TV operator.

The telcos have a strong interest in staying with the star architecture. For years, Raychem Corporation's Raynet unit tried unsuccessfully to get them to consider tree and branch structures. Bellcore said no, and the RBOCs universally said no. Switching at the telcos will probably be a combination of today's digital switches, which themselves will migrate to cell switching or ATM switching fabrics, and the placement of stand-alone ATM boxes.

Oracle Systems expects to have a box on the market next year that could simultaneously service 66,000 homes, all of which might want to watch the same movie at the same time. To be able to funnel programming into those pipes, flexible switching architectures, such as ATM, are needed in front of them, and then the telco must widen the wire. This service could be accomplished several different ways. Companies could use a combination of digital compression and ADSL, which would provide each home with a single channel running at 1.5 megabytes. On a cost-effective investment, they could immediately offer pay-per-view movie service, but they have only one channel per home. Most U.S. homes have 2–4 television sets, so everybody will have to watch the same movie simultaneously because it will be the only service coming into the home.

The more elaborate architectures are those with broadband technologies, offered by AT&T and others, that bring fiber to a distribution point outside the home (fiber to the curb) and then use a coaxial cable dropped into the home. Often the switches and fiber are not provided by the same company. The traditional digital companies perform the telephone function and use some ATM–video server combination to employ the video jukebox, which allows subscribers to select from a broad range of titles and effectively receive video on demand.

For cable, where the capacity exists but the interactivity does not, bidirectional fiber amplifiers are placed on existing coaxial cable or distribution trunks and branches to increase the pay-per-view feed on a return channel. For example, the digital video compression units can take 5 channels of a 30-channel cable system and digitize those 5 channels to provide 25 incremental channels. Companies can run five movies on five channels each starting ten minutes apart to give the appearance of video on demand. They can use the small return channel to retrieve the selections. Cable plant will not offer true interactivity until 1996–98. It will be overbuilt, much like telco overbuilds in the United Kingdom. The switching they will use includes ATM switches from Newbridge. Both architectures will use a common set of switching elements.

Question: Please provide a framework for evaluating competing technologies. How do you decide who has the best mousetrap in a particular area?

Lewis: Discerning who has the best mousetrap is difficult, especially for those who lack a physics or engineering background. What analysts do with the numbers does not matter if they missed the softer, fundamental issues. The company may not grow 30 percent annually if it does not have the mousetrap the analyst thought it had. To get the help needed to make such distinctions, analysts and investors can read trade journals and attend trade shows, visiting the vendor booths and asking customers and suppliers to differentiate the products. Most importantly, they can take questions to the customers themselves. They can ask them what they are looking for, the equipment, the cost bogey, and who is meeting those bogeys and performance standards.

Watching the fishing licenses go out, following which companies win the right to supply various contracts, and watching as those licenses are transformed into real orders and backlog can also be beneficial. The process gets easier with experience, but it is a daunting task the first time out. Investors should start with the larger suppliers and get to know their product lines. They can then balance the smaller suppliers against the larger ones and begin to note the competitive differences.

Question: What companies outside the United States will offer meaningful competition to U.S. equipment manufacturers?

Lewis: Ericsson already provides meaningful competition with an analog cellular market share that includes 40 percent of the world's subscribers. That breaks down to a 35 percent share in the United States and a 70 percent share in Europe. When it moves to digital technology, Ericsson will take the lion's share of the most popular standards, which are TDMA based. Siemens has enjoyed limited success in the central office equipment market, and they will build modestly on that success. Alcatel obtained a domestic presence when it bought the Rockwell transmission unit, and it is now expecting to expand on that foothold.

These suppliers are the preeminent ones globally, which is very favorable for the investment dynamics of this industry and has helped the trade balance. General Instrument and Scientific–Atlanta meet one another more often in Indonesia than they meet any other competitor. This trend is not abundantly clear from the raw trade numbers, because sectors such as customer premises equipment and consumer phones include a lot of equipment imported from the Far East, which distorts the numbers. Network equipment exports outpace imports. The most competitive threat is from Ericsson, followed by Alcatel and Siemens.

Glossary[1]

Analog. The standard method of transmitting a radio phone call. The call is converted into electrical impulses that travel in the form of radio waves, which are analogs (analogous) to the sound waves of the original voice.

Asynchronous digital subscriber line (ADSL). A compression technology that allows twisted pairs of copper lines to carry video signals.

Asynchronous transfer mode (ATM). Used to describe a public high-speed data-transmission system that does not require network synchronization at the customer's premises.

Bandwidth. A range of radio frequencies occupied by a modulated carrier wave assigned to a service.

Broadband. Of, having, or involving operation with uniform efficiency over a wide band of frequencies.

Build-out. The process of implementing a telecommunications service network.

Bypass. A method of communication that avoids accessing the local-exchange carrier.

Cable television (CATV). Television transmitted over cable into a building.

Call forwarding. The process by which telephone calls to one number can be automatically sent to or forwarded to another.

Cell. The basic geographical unit that gave cellular its name. A cellular network in a city or county is carved into "cells," each of which is equipped with a radio transmitter/receiver. A cell's size depends on terrain and capacity demands. A computer at the network's switching office monitors each call, handing it off to another cell and at another frequency as needed. (See **handoff, microcell.**)

Cellular digital packet data (CDPD). A method of transmitting data over the cellular network that places the message in digital electronic "envelopes" and sends it at high speed through underused radio channels or between pauses in cellular-phone conversations. CDPD, which can send data as much as eight times faster than regular cellular delivery, is being introduced in some major markets in 1994.

Channel. The width of a spectrum band, measured in kilohertz, taken up by a radio signal. Most phones use 30-kilohertz channels. (See **hertz, spectrum.**)

---

[1] Portions reprinted by permission of *The Wall Street Journal*, copyright ©1994, Dow Jones & Company, Inc. All rights reserved worldwide.

Coaxial cable. A line used to transmit telegraph, telephone, and television signals of high frequency.

Code division multiple access (CDMA). A type of digital-transmission system said to offer up to 20 times more capacity than conventional analog and digital cellular systems. CDMA systems use a low-power signal spread across a wide-frequency band. The systems assign a special electronic code (instead of a frequency) to each call signal. They thus increase the number of calls that can occupy the same space in a communications channel and allow them to be spread over an entire frequency band. (See also **time division multiple access**.)

Competitive access provider (CAP). A company that provides access to long-distance carriers in competition with local-exchange companies.

Consent decree. Refers to the original 1956 consent decree that led to the breakup of the former "Bell System."

Customer's premises equipment (CPE). A term used to describe any telecom device located within a customer's premises (as distinguished from devices in a service company's outside plants or central office).

Digital. The transmission method for sending voice or information using the binary code (ones and zeros) of computer language. It is accomplished by transmitting on/off electrical pulses.

Direct broadcast satellite (DBS). A broadcast service authorized by the FCC that would broadcast television signals directly to small individual receiving stations located at private buildings.

Enhanced specialized mobile radio (ESMR). Paging, dispatch, and cellular services offered over cellularized dispatch frequencies.

Federal Communications Commission (FCC). The regulatory commission established by the Communications Act of 1934 to administer interstate common carriage of telecommunications. The five commissioners serve staggered terms and are appointed by the President when vacancies occur, with confirmation by Congress. The chair is designated by the President and serves at the President's pleasure. No more than three of the commissioners may be from the same political party.

Frequency reuse. The ability to use the same frequencies repeatedly across the different cells of a cellular system, thereby increasing call-handling capacity despite a limited number of channels. Frequency use is made possible by the basic cellular setup; that is, each cell uses the radio frequencies within its boundaries, which allows the same frequencies to be reused with little or no interference from nearby cells.

Gigahertz (GHz). A unit of frequency equal to one billion hertz.

Handoff. The process by which a switching office passes a cellular conversation from one radio frequency in one cell to another frequency in another cell.

Handset. A combined telephone transmitter and receiver mounted on a handle.

Head end. Distribution center for a cable delivery system.

Hertz. Cycles per second, the international measure of frequency. When used in relation to the radio spectrum, hertz is generally preceded by prefixes, as in kilohertz (thousands of cycles per second) and megahertz (millions of cycles). (See **spectrum**.)

Home shopping service. A television program during which various goods are advertised and sold to viewers, who place orders by telephone.

Interactive. Two-way communication that allows viewers to select educational, informational, entertainment, or shopping options.

Interexchange carrier (IXC). A long-distance carrier.

Landline. Wired; a landline telephone service is a wired phone service.

Lifeline Residential Services. Telephone services provided to low-income residents, sometimes at deep discounts from regular rates.

Local-exchange company (LEC). A company providing local telephone services, including access to long-distance carriers.

Megahertz (MHz). A unit of frequency equal to one million hertz.

Metropolitan statistical area (MSA). One of the 306 largest urban population markets as defined by the U.S. government; each MSA has two licensed cellular-service operators.

Microcell. A unit that is smaller than a cell and will be the future unit into which personal communication service (PCS) networks will be divided. PCS microcellular networks will use thousands of tiny radio transceivers in myriad microcells. (See **personal communication service**.)

Minutes of use (MOU). A unit of measure to describe telephone use by time.

Mobile telephone switching office (MTSO). The central computerized switch that controls the entire operations of a cellular system—monitoring all cellular calls, tracking the location of all cellular-equipped vehicles traveling in the system, arranging handoffs, and billing.

Modification of Final Judgment (MFJ). Refers to the modification of the consent decree that led to the breakup of the former "Bell System" in the United States.

Multichannel Multipoint Distribution System (MMDS). More commonly known as wireless cable.

Multimedia. The blending of different types of media—audio, video, data—into one product or service.

Negroponte switch. Theory, launched by Nicholas Negroponte of MIT's Media Lab, that what is currently transmitted by air (mainly broadcast video) would switch to fiber optics and coaxial wires while what is currently transmitted by wire (mainly voice telephony) would move to air.

Nonwireline. A cellular-phone carrier, one of two in any market, that is independent of the regular local phone service. Customers who use the nonwireline cellular carrier operate on the cellular network's A-band. (See **wireline**.)

On-demand entertainment. Video delivered to the customer's home at the customer's demand through telephone, cable, or wireless transmission.

On-line system. A system or program that is electronically hooked up to a telephone line.

Pay-per-view television. Television programs that charge additional fees when the customer orders the program broadcast.

Personal communication network (PCN). A wireless telecommunications service. (See **personal communication service**.)

Personal communication service (PCS). Digital wireless service that will let users make calls from lightweight pocket phones or send data messages from tiny communicators called personal digital assistants. Expected to become widely available between 1997 and 1999, PCS systems will use microcells and operate at higher frequencies and in a broader slice of the spectrum than cellular systems.

Personal digital assistant (PDA). Small, hand-held data communicator that can send messages over wireless networks.

Plain old telephone service (POTS). Basic, voice-grade, switched telephone service.

Point of presence (POP). Location where a long-distance carrier has installed transmission and/or switching equipment in a service area that serves as, or relays to, a network switching center of that long-distance carrier.

Pop. A term for the population in an area served by a carrier. An area with one million people is said to have one million pops—a million potential customers.

Private branch exchange (PBX). Telephone switching equipment located on a customer's premises and dedicated to use by that organization.

Public utility commission (PUC). A state regulatory authority set up to monitor utility monopolies such as local telephone services.

Regional Bell operating company (RBOC). A local-exchange business of the former "Bell System." The seven RBOCs are Ameritech, NYNEX, Pacific Telesis, Bell Atlantic, SouthWestern Bell, U S West, and BellSouth.

Roaming. Refers to a cellular-phone subscriber's use of a phone beyond his or her usual service area, such as when driving from one state to another.

Rural service area (RSA). One of 428 designated geographical regions in the United States and its territories. The FCC awarded two cellular licenses to each RSA.

Service plan. A rate plan that a subscriber chooses when requesting cellular service. It usually consists of a base rate for connecting to the system and a per-minute charge for use.

Specialized mobile radio (SMR). A two-way radio dispatch service operating in the 800–900-megahertz frequency band. Following the recent relaxation of federal restrictions, SMR operators who had previously handled taxi dispatches have expanded into cellularlike phone services, which has put them in direct competition with cellular-phone companies. (See **enhanced specialized mobile radio.**)

Spectrum. The range of electromagnetic waves, which vary in length (or frequency) and intensity, used in wireless communications. The range encompasses sound near the low end, light toward the high end, and in between, the radio spectrum. Spectrum is also used to refer simply to radio wavelengths.

Telco. Shortened version of telephone company.

Telephony. The use or operation of an apparatus for transmission of sounds between widely removed points.

Time division multiple access (TDMA). A digital-transmission scheme for expanding the capacity of cellular systems. TDMA cannot increase capacity as much as the CDMA approach, but the TDMA technology is further developed. Now being installed by many carriers, TDMA transmission equipment offers three-to-seven times the capacity of existing cellular analog systems. Current TDMA systems divide a channel into three time slots, each a fraction of a second, which allows one channel to handle three calls simultaneously. Future systems will allow seven calls at once.

Video dial tone. Refers to the use of the telephone system to distribute video.

Video-on-demand service. A service that can deliver any program on demand—that is, at any time—and thus allows viewers to control programming.

Wireless. Refers to nonwired telecommunications such as cellular.

Wireline. A cellular-phone carrier owned by the company that also operates the regular local phone service in a given market. Customers who use the wireline cellular carrier operate on the cellular network's B-band.

Order Form

Additional copies of *The Telecommunications Industry* (and other AIMR publications listed on page 112) are available for purchase. The price is **$20 each in U.S. dollars**. Simply complete this form and return it via mail or fax to:

AIMR
Publications Sales Department
P.O. Box 7947
Charlottesville, Va. 22906
U.S.A.
Telephone: 804/980-3647
Fax: 804/977-0350

Name _____

Company _____

Address _____

_____ Suite/Floor _____

City _____

State _____ ZIP _____ Country _____

Daytime Telephone _____

Title of Publication   Price   Qty.   Total

_____   _____   _____   _____

_____   _____   _____   _____

Shipping/Handling
- ❏ All U.S. orders: Included in price of book
- ❏ Airmail, Canada and Mexico: $5 per book
- ❏ Surface mail, Canada and Mexico: $3 per book
- ❏ Airmail, all other countries: $8 per book
- ❏ Surface mail, all other countries: $6 per book

Discounts
- ❏ Students, professors, university libraries: 25%
- ❏ CFA candidates (ID #_____):   25%
- ❏ Retired members (ID #_____):   25%
- ❏ Volume orders (50+ books of same title):   40%

Discount $ —_____

4.5% sales tax
(Virginia residents) $ _____

8.25% sales tax
(New York residents) $ _____

7% GST
(Canada residents,
#124134602) $ _____

Shipping/handling $ _____

Total cost of order $ _____

❏ Check or money order enclosed payable to **AIMR**   ❏ Bill me
Charge to: ❏ VISA   ❏ MASTERCARD   ❏ AMERICAN EXPRESS

Card Number:_____ ❏ Corporate ❏ Personal

Signature:_____ Expiration date: _____

Selected AIMR Publications*

Managed Futures and Their Role in Investment Portfolios, 1994 . . . . $20
  Don M. Chance, CFA

Fundamentals of Cross-Border Investment: The European View, 1994 . . . . $20
  Bruno Solnik

Good Ethics: The Essential Element of a Firm's Success, 1994 . . . . $20
  H. Kent Baker, CFA, *Editor*

A Practitioner's Guide to Factor Models, 1994 . . . . $20

Quality Management and Institutional Investing, 1994 . . . . $20
  Keith P. Ambachtsheer, *Editor*

Managing Emerging Market Portfolios, 1994 . . . . $20
  John W. Peavy III, CFA, *Editor*

Global Asset Management and Performance Attribution, 1994 . . . . $20
  Denis S. Karnosky, Ph.D., and Brian D. Singer, CFA

Franchise Value and the Price/Earnings Ratio, 1994 . . . . $20
  Martin L. Leibowitz and Stanley Kogelman

Investing Worldwide, 1993, 1992, 1991, 1990 . . . . $20 each

The Modern Role of Bond Covenants, 1994 . . . . $20
  Ileen B. Malitz

Derivative Strategies for Managing Portfolio Risk, 1993 . . . . $20
  Keith C. Brown, CFA, *Editor*

Equity Securities Analysis and Evaluation, 1993 . . . . $20

The CAPM Controversy: Policy and Strategy Implications for
  Investment Management, 1993 . . . . $20
  Diana R. Harrington and Robert A. Korajczyk, *Editors*

The Health Care Industry, 1993 . . . . $20
  James Balog, *Editor*

Predictable Time-Varying Components of International Asset Returns, 1993 . . . . $20
  Bruno Solnik

The Oil and Gas Industries, 1993 . . . . $20
  Thomas A. Petrie, CFA, *Editor*

Execution Techniques, True Trading Costs, and the Microstructure
  of Markets, 1993 . . . . $20
  Katrina F. Sherrerd, CFA, *Editor*

Investment Counsel for Private Clients, 1993 . . . . $20
  John W. Peavy III, CFA, *Editor*

Active Currency Management, 1993 . . . . $20
  Murali Ramaswami

---

*A full catalog of publications is available from AIMR, P.O. Box 7947, Charlottesville, Va. 22906; 804/980-3647; fax 804/977-0350.